Snow, Ice and
Other Wonders of
Water
A Tribute to the
Hydrogen Bond

Snow, Ice and
Other Wonders of
Water

A Tribute to the
Hydrogen Bond

Ivar Olovsson

University of Uppsala, Sweden

World Scientific

NEW JERSEY · LONDON · SINGAPORE · BEIJING · SHANGHAI · HONG KONG · TAIPEI · CHENNAI · TOKYO

Published by

World Scientific Publishing Co. Pte. Ltd.

5 Toh Tuck Link, Singapore 596224

USA office: 27 Warren Street, Suite 401-402, Hackensack, NJ 07601

UK office: 57 Shelton Street, Covent Garden, London WC2H 9HE

Library of Congress Cataloging-in-Publication Data

Names: Olovsson, Ivar, 1928- author.

Title: Snow, ice and other wonders of water : a tribute to the hydrogen bond /
Ivar Olovsson, University of Uppsala, Sweden.

Description: New Jersey : World Scientific, [2016] | Includes index.

Identifiers: LCCN 2015046021| ISBN 9789814749350 (hardcover : alk. paper) |
ISBN 9814749354 (hardcover : alk. paper) | ISBN 9789814749367 (pbk. : alk. paper) |
ISBN 9814749362 (pbk. : alk. paper)

Subjects: LCSH: Snow--Popular works. | Water--Popular works. | Ice crystals--Popular works. |
Water chemistry--Popular works. | Hydrogen bonding--Popular works. |
Phase transformations (Statistical physics)--Popular works.

Classification: LCC QC926.32 .O46 2016 | DDC 551.57/84--dc23

LC record available at http://lccn.loc.gov/2015046021

British Library Cataloguing-in-Publication Data

A catalogue record for this book is available from the British Library.

Preface

Water plays a fundamental role in life and, although a tremendous amount of research has been performed on this seemingly simple substance, there are still many unresolved questions. The literature is virtually unlimited and it is impossible to cover more than a few topics in this book. I have chosen to concentrate on snow and ice, and I hope the book will whet the appetite to learn more about this fascinating field. The selected topics are mostly rather briefly treated but further information is easily available on the Internet. It therefore seems unnecessary to include a long list of references here.

Water plays a unique role in chemistry. The special properties of the different forms of water — from ice and snow to liquid water — are due to hydrogen bonding between the H_2O molecules, and this book is a tribute to a field to which I have spent a major part of my research. The hydrogen bond is of fundamental importance in biological systems since all living matter has evolved from and exists in an aqueous environment. Hydrogen bonds are involved in most biological processes as little energy is needed in forming as well as breaking these bonds.

I wish to express my sincere thanks to Prof. Yoshinori Furukawa of the Institute of Low Temperature Science (Hokkaido University), British artist Simon Beck, Canadian photographer Don Komarechka, and Russian photographers Alexey Kljatov and Andrei Osokin for their kind permission to include their beautiful and sometimes unique pictures. I hope the pictures reproduced here will stimulate further studies regarding their books.

I am grateful for permission from publisher De Gruyter to reprint my articles from *Zeitschrift für Physikalische Chemie*, from publisher VCH Verlagsgesellschaft GmbH to reprint a paper from *Angewandte Chemie, International Edition* and from the Harbourcreek Historical Society to reprint an article from their newsletter. My thanks to Prof. Kenneth Libbrecht for permission to use his snow crystal morphology diagram. Many thanks to Prof. Emer. Anders Liljas for checking the manuscript and for his valuable suggestions. I welcome suggestions for improvements and the alerting of errors.

Ivar Olovsson
Uppsala, 15 October 2015
Ivar.Olovsson@gmail.com

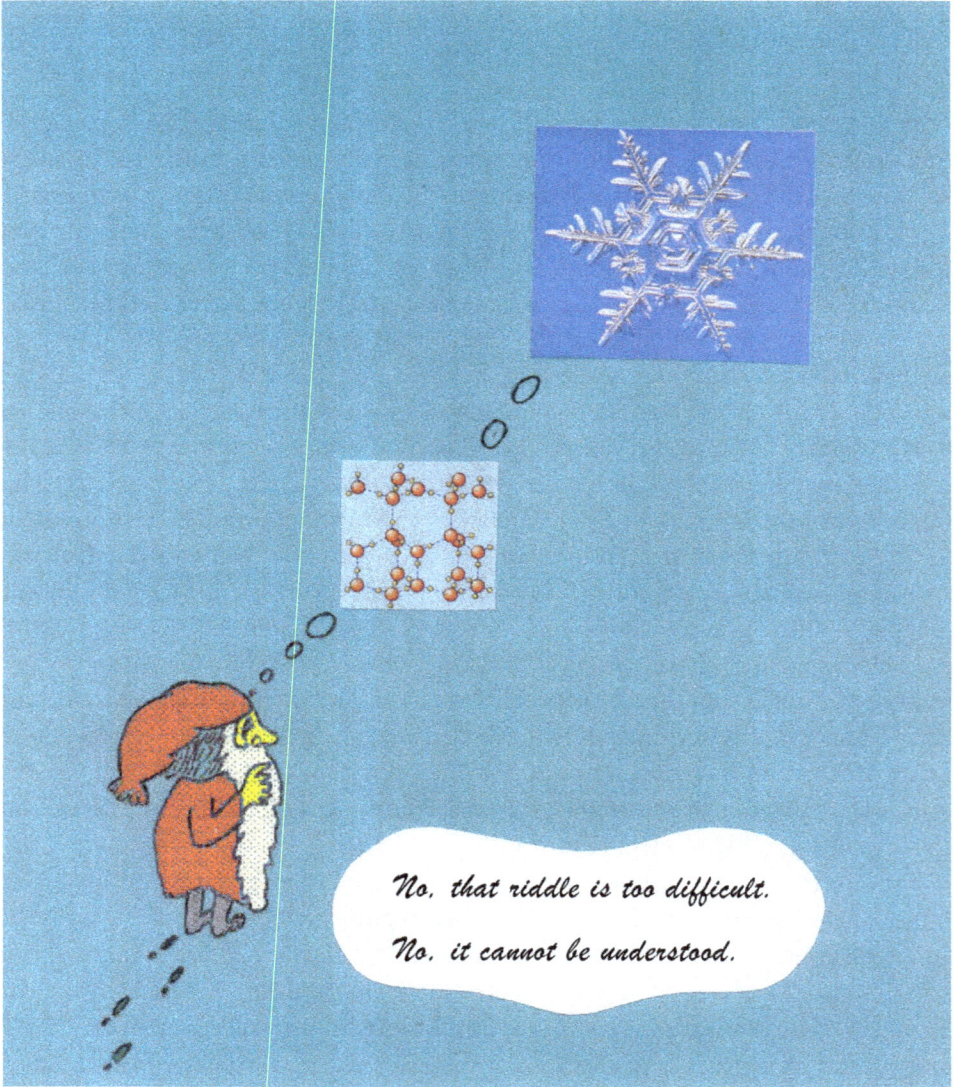

No, that riddle is too difficult.

No, it cannot be understood.

Contents

1 There Are Many Different Types of Snow

Frozen water — snow and ice — appears in a myriad of different shapes and with properties which can be quite different. Detailed knowledge of the properties of snow is of great importance for the Sami people (Laplanders) involved in reindeer herding. A large number of names are used by the Laplanders to characterize the different types, and snow may still be a daily topic of conversation. In Yngve Ryd's book (in Swedish) *Snö: En Renskötare Berättar* (*Snow: A Laplander Narrates*), more than 300 words for "snow" are documented with explanations and photos. The words describe for instance the amount of snow, consistency, gliding, buoyancy or melting.

In this book I will mostly use the term "snow crystal" for a single crystal, i.e. a sample which is continuous and has no boundaries (all parts of the crystal are extinguished simultaneously in polarized light). A fully developed dendritic snow crystal is, for example, a single crystal. All snow crystals (in my terminology) are transparent. As a snow crystal falls toward the earth, it will often hook onto other crystals in a random way and a *snowflake* is formed. A layer of snow looks white owing to repeated reflection of the light toward the randomly oriented snowflakes. In the literature the word "snowflake" seems to be used for all types — a single crystal as well as a random collection of snow crystals. A snow crystal is just ordinary ice, but ice with a special, mostly rather open structure. A large, more compact and irregular crystal is better named *ice crystal* (see photo on the left by Andrei Osokin).

2
Early Snow Crystal Observations

Snow has always fascinated mankind and is mentioned in a dozen places in the Bible. In Job 38:22–23 is written: "Hast thou entered into the treasure house of the snow, or hast thou seen the treasure house of the hail, which I have reserved against the time of trouble, against the day of battle and war?" Hail is often considered to be punishment from God — in Revelation 16:21 is written: "And there fell upon men a great hail out of heaven, every stone about the weight of a talent: and men blasphemed God because of the plague of the hail; for the plague thereof was exceeding great." (One talent was about 50 kg.)

To my knowledge, the first pictures of snow crystals were published in 1555 by Olaus Magnus in his famous *Historia de Gentibus Septentrionalibus* (*History of the Nordic People*, an assembly of essays in 22 volumes; Fig. 2.1). Owing to the Reformation in Sweden, he lived at that time in Rome together with his brother, Catholic archbishop Johannes Magnus. In exile he seems to have largely forgotten how snow crystals look and applied his fantasy.

It is commonly considered that it was the astronomer Johannes Kepler who, in his essay *De Nive Hexangula*, first established that snow crystals have a six-fold symmetry (Figs. 2.2 and 2.3). This conclusion appears to be based on his studies of the closest packing of spheres. In this context he had also been asked to solve a practical problem: how best to stack cannonballs on ships.

Fig. 2.1. Snow crystals drawn by Olaus Magnus.

Fig. 2.2. Johannes Kepler
(1571–1630).

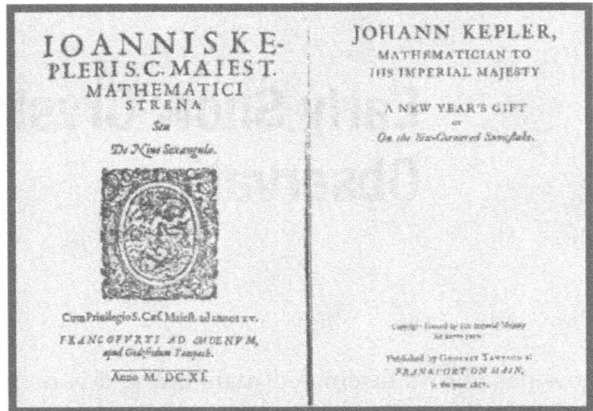

Fig. 2.3. Kepler's book *De Nive Sexangula*
(*On the Six-Cornered Snowflake*), published in 1611.

The French philosopher and mathematician René Descartes (Fig. 2.4) made, during the unusually cold winter in Amsterdam in 1635, the first detailed observations of snow crystals, which were published in 1637 in his famous work *Discours de la Methode* (Fig. 2.5). The pictures in Fig. 2.6 are found in the chapter *"Les Meteores."* This may seem strange in a treatise dealing with snow, but *Meteorology* deals with all atmospheric phenomena, such as wind, storms, cyclones, rain, snow and hail.

Descartes was born at La Haye in Touraine, France. The village is nowadays named "Descartes," in his honor. In 1649 he was invited to Stockholm by Queen Kristina to be her teacher and adviser, and to organize a new scientific academy

Fig. 2.4. René Descartes (1596–1650).

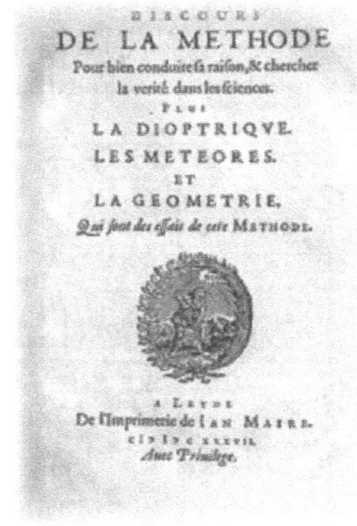

Fig. 2.5. *Discours de la Methode* (1637).

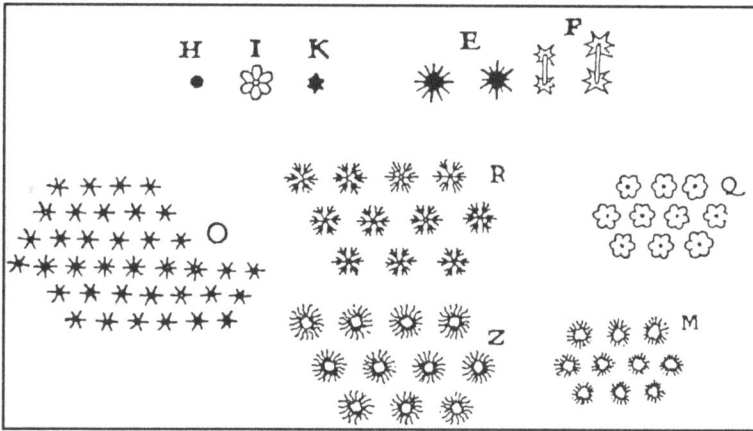

Fig. 2.6. Snow crystals observed by Descartes.

Fig. 2.7. Descartes and Queen Kristina.

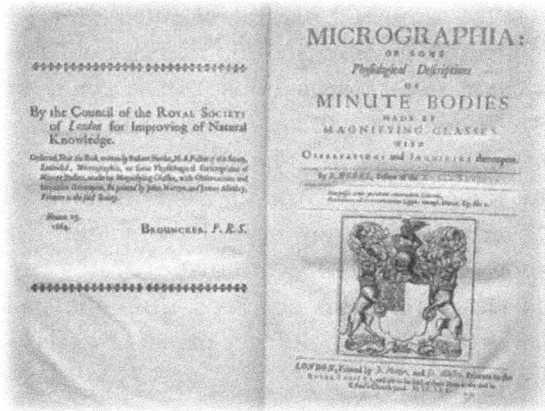

Fig. 2.8. Robert Hooke's *Micrographia*, published in 1665.

(Fig. 2.7). It has been said that Descartes considered Sweden a country where both people and thoughts froze to ice. The meetings were held early in the morning (at 5 a.m.) in the castle which was hardly heated and very cold and draughty. Descartes caught a cold and died of pneumonia on February 11, 1650, after only a few months in Sweden.

Many other scientists have also pondered on the mysteries of snow crystals. When the microscope was invented in the later part of 1600, the possibilities of studying snow crystals became much better. In 1665 the multidimensional scientist and polymath Robert Hooke published *Micrographia: Or Some Physiological Descriptions of Minute Bodies Made by Magnifying Glasses with Observations and Inquiries Thereupon* (Fig. 2.8). An imagined picture of Hooke at his desk is shown in Fig. 2.9 (no contemporary portrait has been preserved). A few of his drawings of snow crystals are shown in Fig. 2.10. Hooke remarked that the angle between the side branches is always 60°.

Fig. 2.9. Robert Hooke (1635–1703).

Fig. 2.10. Snow crystals drawn by Robert Hooke.

The eminent natural scientist and whaler William Scoresby, Jr. (Fig. 2.11) published in 1820 a two-volume book, *An Account of the Arctic Regions with a History and Description of the Northern Whale Fishery* (Fig. 2.12). In this famous work he also made accurate observations of snow crystals, some of which are shown in Fig. 2.13. Note that Scoresby (as well as Descartes) also observed three-dimensional snow crystals: two or three planar crystals joined by an axis.

Fig. 2.11. William Scoresby, Jr. (1789–1857).

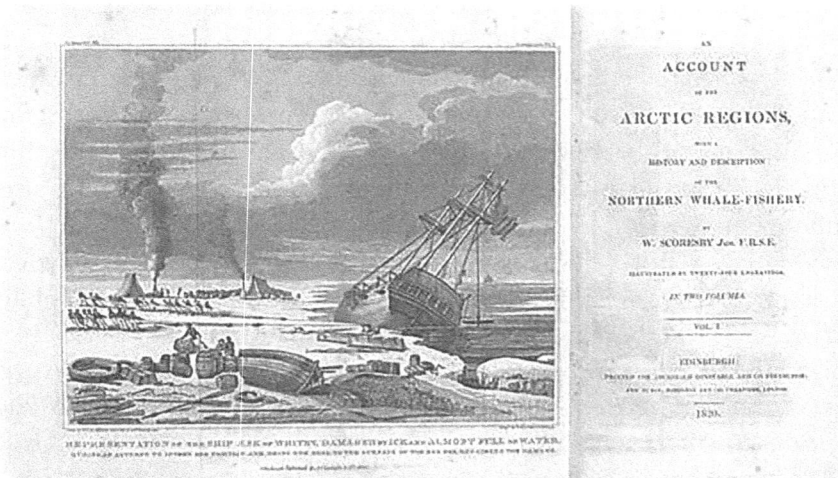

Fig. 2.12. An Account of the Arctic Regions with a History and Description of the Northern Whale Fishery.

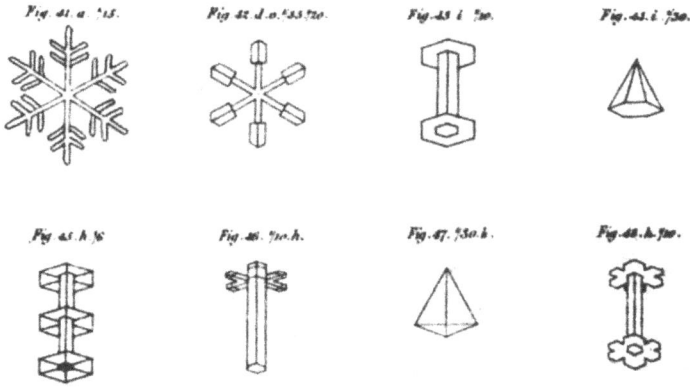

Fig. 2.13. Snow crystals observed by William Scoresby, Jr.

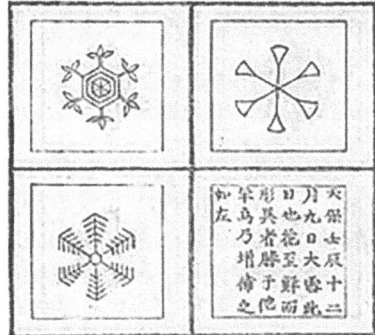

Fig. 2.14. Snow crystals from the book *Sekka Zusetsu* by Doi Toshitsura.
(From an exhibit in the National Museum of Nature and Science, Tokyo, Japan.)

The *daimyo* (Japanese feudal lord) Doi Toshitsura wrote in 1832 the book *Sekka Zusetsu*, on snow crystals (Fig. 2.14). Note that in one of the pictures the branches are directed inward — a most unlikely situation, and probably never observed.

The English meteorologist and aeronaut James Glaisher (Fig. 2.15) published in 1855 a collection of snow crystals: *Photogenic Drawings of Snow Crystals, as Seen in January 1854.* The

Fig. 2.15. James Glaisher (1809–1903).

Fig. 2.16. Glaisher's drawings of snow crystals.

Fig. 2.17. Wilson A. Bentley (1865–1931).

Fig. 2.18. Snow crystals by Bentley.

snow crystals shown in Fig. 2.16 were sketched by Glaisher and drawn properly by his wife, artist Cecilia Louisa Glaisher. These drawings of snow crystals are considered to be among the best ever published.

Wilson "Snowflake" Bentley from Jericho, Vermont, USA, was one the first known photographers of snowflakes (Fig. 2.17). He was a farmer without a scientific background. He attached a bellows camera to a compound microscope and, after much experimentation, photographed his first snowflake on January 15, 1885. For almost half a century Bentley captured and photographed more than 5,000 snowflakes. He poetically described snowflakes as "tiny miracles of beauty." He wanted them to appear "like diamonds on velvet," so he carefully cut the photographs and mounted them on black paper. His pictures were spread all over the world and published in many leading magazines. In 1931 the American Meteorological Society gathered the best of his pictures in a monograph illustrated with 2,500 pictures; a few are shown in Fig. 2.18. Bentley died of pneumonia at his farm on December 23, 1931, after walking home six miles in a blizzard. His book *Snow Crystals* was published shortly before his death.

3
Artificial Snow Crystals

Wilson A. Bentley's beautiful snow pictures inspired many scientists and artists, among them the Japanese nuclear physicist Ukichiro Nakaya (Fig. 3.1). In 1932 Nakaya got a position in the newly established science faculty at the university in Sapporo on the island of Hokkaido in northern Japan. No apparatus for research in nuclear physics was available there, so he directed his interest toward research material which was unlimited around Sapporo — *snow*. During a series of winters he made careful studies of snow crystals in the mountains around Sapporo and found that regular hexagonal crystals were not as common as more irregular ones. His findings and classification are summarized in Fig. 3.2. (New classification schemes have later been introduced. The widely used Mogano–Lee scheme from 1966 contains 80 types. K. Kikuchi, T. Kameda, K. Higushi and A. Yamashita have in 2013 suggested a global classification scheme with 121 types.)

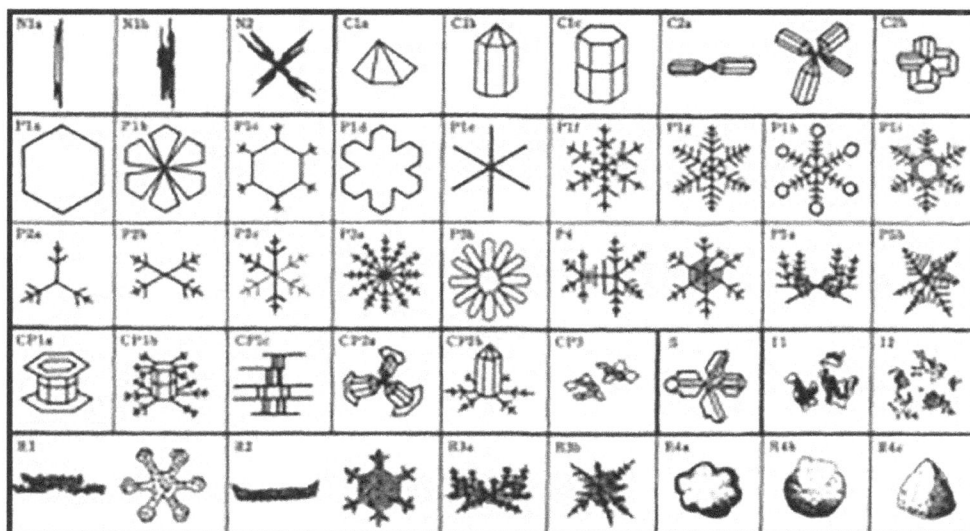

Fig. 3.1. Ukichiro Nakaya (1900–1962).

Fig. 3.2. Different snow crystal forms found by Nakaya.

Nakaya wondered why the crystals were so different — what were the atmospheric conditions when these different types of snow crystals were formed high up in the air? As it was not possible to determine the exact meteorological situation for each individual crystal, he decided to build a laboratory where he could grow snow crystals under different temperature and humidity conditions. In 1935 the low-temperature laboratory was opened. No one had previously made artificial snow crystals and it took a considerable time before Nakaya found a good method. Attempts to grow the crystals on cold surfaces only resulted in typical frost patterns. Finally, he succeeded in growing real snow crystals on rabbit hair in cold humid air, and on March 12, 1936 he created the first artificial snow crystal. Rabbit hair has small knobs at suitable distances and these are evidently suitable starting points for crystals to grow. It took 30–60 minutes before a typical branchy, dendritic snow crystal was formed.

Owing to the Second World War, it took a long time before Nakaya's collected works were published. The original text and many pictures were destroyed when the printing house was bombed. In 1954 the book *Snow Crystals: Natural and Artificial* was published. This is beautifully illustrated and summarizes Nakaya's research on snowflake crystals, starting from his work at Hokkaido University. The original book has long been out of print but reprints are available. It serves as a classic reference on crystal shapes, showing how a scientific investigation can proceed through systematic observation toward an accurate description of a fascinating natural phenomenon.

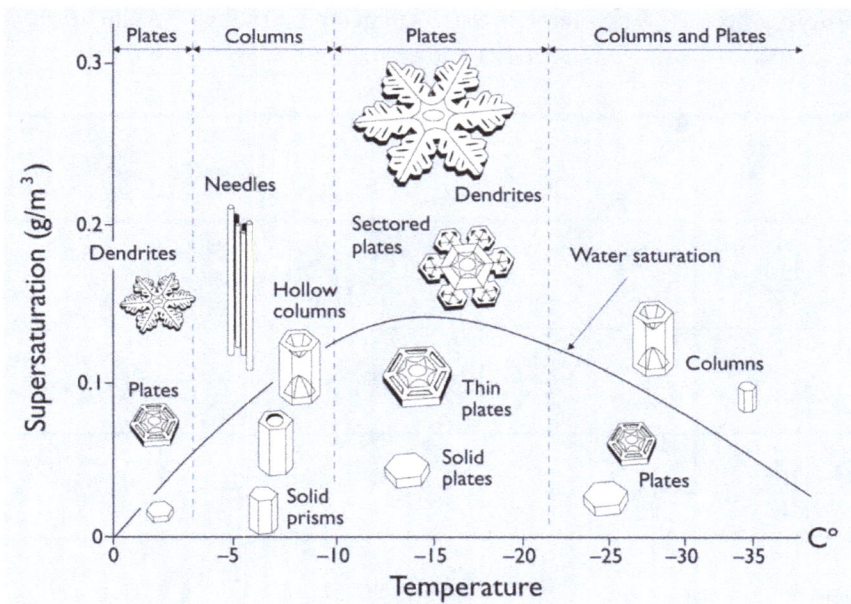

Fig. 3.3. Dependence of crystal shape on temperature and humidity.
(© Kenneth Libbrecht, permission granted.)

The "Nakaya" diagram in Fig. 3.3 displays the influence of temperature and humidity on the crystal shape. The vertical axis shows the density of water vapor in the excess of saturation with respect to ice. The curved line shows the saturation with respect to liquid water as a function of temperature. The borders between the different temperature ranges are very sharp — less than 1°C. Plates and prisms with planar surfaces are formed at low humidity. As the humidity increases, the edges and corners grow fastest and cavities are formed (the addition of water molecules from the environment and the disposal of the heat of crystallization is most effective at the edges and corners). As the humidity increases further, snow crystals with broad points are first formed, and these points gradually become narrower, and finally side branches are developed, resulting in so-called dendritic crystals at the highest humidity.

As a growing crystal is moved from one environment to another with different temperature or humidity, a mixed crystal will be formed, like those shown in Fig. 3.2. This is just what happens as a natural snow crystal encounters different atmospheric conditions when it falls to the earth. The resulting snow crystal is unique for its meteorological history; "the snow crystal is its own tachometer." Or, as Nakaya expressed it: "Snow crystals are the hieroglyphs sent from the sky."

At the low-temperature laboratory in Sapporo, snow research has continued under the leadership of Teisaku Kobayashi (Fig. 3.4) and present head Yoshinori Furukawa. A few examples of snow crystals grown in the laboratory are shown in Figs. 3.5a–f. The pictures may serve as an illustration of the gradual development of crystals with increasing humidity. The crystal with 12 branches appears to be a twin (*cf.* Descartes' pictures in Fig. 2.6).

Fig. 3.4. Teisaku Kobayashi at work growing snow crystals.

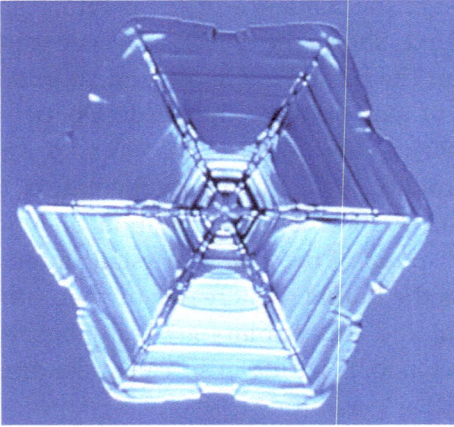

Fig. 3.5a

Fig. 3.5b

Fig. 3.5c

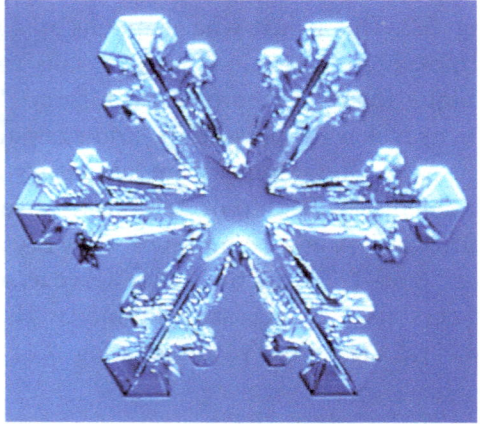

Fig. 3.5d

Fig. 3.5e

Fig. 3.5f

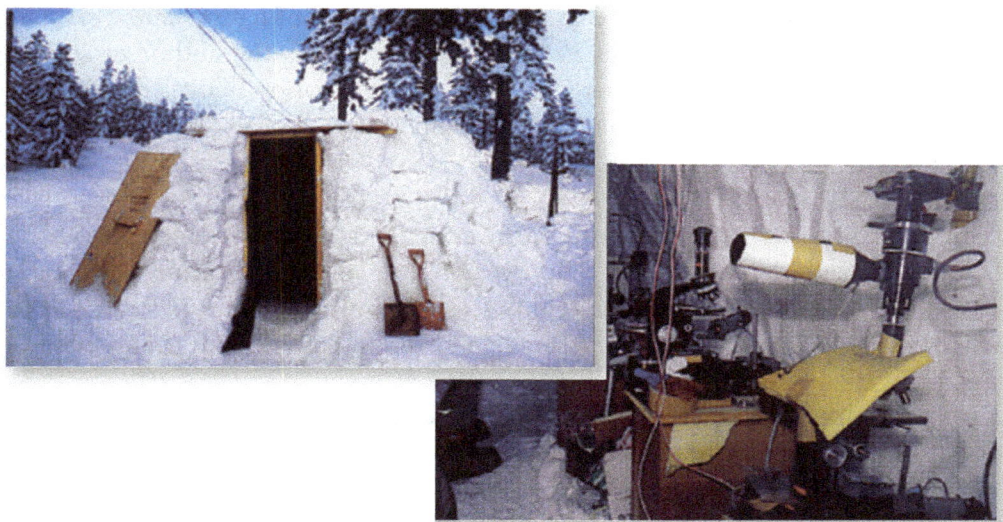

Fig. 3.6. Furukawa's snow hut for studying natural snow crystals.

Besides work in the low-temperature laboratory, Yoshinori Furukawa has followed up Nakaya's old field studies — taking pictures of natural snow crystals in a specially equipped "snow hut" (Fig. 3.6). Under the same conditions, the natural and artificial snow crystals are indistinguishable. Furukawa is in particular involved in experimental and theoretical studies of crystal growth and various surface phenomena. A comprehensive summary of his work appears in *Handbook of Crystal Growth* (Elsevier, 2015), Vol. I, pp. 1061–1112.

A neat way of studying snow crystals is illustrated below (Fig. 3.7). Don Komarechka has taken a large number of very beautiful macro photos in nature, some of which are reproduced below. A collection of his photographs is shown in the recent book *Sky Crystals: Unraveling the Mysteries of Snowflakes*. Exposure details and explanations of special features are also provided in the book.

Scientific studies of snow and ice are today a very active field in many countries. Caltech physicist Kenneth Libbrecht has since childhood been fascinated by the beauty of snow crystals and has traveled the world studying and documenting

Fig. 3.7. Study of snow crystals by Don Komarechka.

them. He is also actively involved in taking photographs of natural snow crystals and in growing artificial crystals in the laboratory. He is the author of several books in the field, and a very large amount of information about the work by him and others on snow crystals is on the Internet: www.caltech.edu/~atomic/snowcrystals.

"Snow crystals" with seven and eight arms.   In Christmas decorations one often finds "snow" crystals with four, five, seven or even eight arms. People are of course free to choose any shape in the decoration — but if it is supposed to illustrate a snow crystal the only acceptable form is a pattern with six branches!

Can you find two identical snow crystals?   The answer to this classical question depends first of all on how closely you are able to look at the crystal. Using the naked eye or a microscope with optical resolution, it seems quite possible that you will find two crystals which you judge to be alike at this level of observation. However, at atomic resolution the answer will be quite different. It is easy to demonstrate that the probability must be extremely small that two crystals are built in exactly the same way. If we assume that a snow crystal weighs 10^{-4} g, such a crystal will consist of around 10^{18} water molecules. If an additional water molecule attaches, it will have around 10^{18} alternative positions to choose from (assuming a flat dendritic snow crystal and neglecting space problems). The probability that the next water molecule will attach to exactly the same place of an imagined identical snow crystal is clearly very small. Furthermore, the water molecules building up the ice structure may have different isotope compositions (with ^{16}O, ^{17}O, ^{18}O, ^{1}H or ^{2}D), which will lead to even more alternative compositions of the ice, and formally non-identical crystals.

Von Koch's snowflake.   This is a purely mathematical construction but the shape has some similarity to natural snow flakes. It was one of the earliest fractal figures described and amongst the most important objects used by B. Mandelbrot for his pioneering work on fractals. It appeared in a 1904 paper by the Swedish mathematician Helge von Koch (Fig. 3.8), *"Sur une courbe continue sans tangente, obtenue par une construction géométrique élémentaire."*

The von Koch snowflake is obtained by trisecting each side of an equilateral triangle and replacing the straight

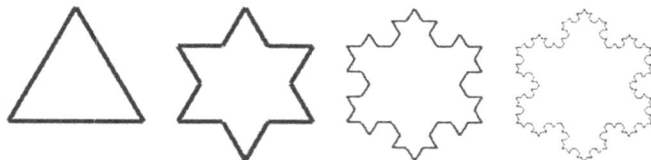

Fig. 3.8. Helge von Koch and the construction of his "snowflake."

segment by two sides of a smaller equilateral triangle projecting outward, then treating the resulting figure the same way, and so on. The figure is self-similar, like all true fractals. The area is limited but the total length of the fully developed curve is infinite.

It seems that the term "fractal" is often used to describe richly branched objects in general, like dendritic structures in plants and trees. Should a snow crystal be characterized as fractal? If we assume that the tiny "sprouts" on the side branches of a fully developed dendritic snow crystal continue to grow in the same way as the big crystal and this procedure is repeated indefinitely, we could perhaps call it fractal. But you will never see such a fully developed snow crystal in practice, so dendritic snow crystals should probably not be characterized as fractal.

Ice patterns on windows.   When water vapor is deposited on a cold surface, such as a glass window, the feather-like patterns developed are quite different from natural snow crystals (Fig. 3.9). It is clear that scratches and impurities are to some extent responsible for the formation of these strange patterns. However, it is difficult to understand why an ice germ started at a certain point does not continue straight ahead from there. Free ice crystals grow normally in certain directions determined by the internal crystal structure. However, it appears that temperature and air humidity play an important role here. If the air is relatively dry and the surface very cold (down to −25°C), a pattern more closely resembling a snow crystal may develop. In air of higher humidity the surface is bombarded with such an amount of water molecules that a typical ice pattern does not have time to develop. Similar patterns may also form on a wooden deck. Here, the probability is larger that feather-like patterns will form on fat (hydrophobic) surfaces.

Twins, snowflakes and hail.   The snow crystals described so far are single crystals, i.e. the entire sample is continuous and there are no grain boundaries.

Fig. 3.9. Ice formation on cold surfaces.

A fully developed dendritic snow crystal is also a single crystal. However, if the growth of the crystal is disturbed, another crystal may develop with another orientation. In some cases such an outgrowth may have a direction which is related to the original structure and a twin or triplet is formed (see Fig. 3.5f).

As the snow crystal falls toward the earth, it will often hook onto other crystals in a random way and a *snowflake* is formed. Very large flakes are especially formed close to 0°C. A single snow crystal is transparent but a layer of snow looks white owing to repeated reflection of the light toward the randomly oriented snowflakes.

If a supercooled droplet of water larger than 0.01 mm comes into contact with an ice crystal, it may not have enough time to spread out over the crystal and result in a regular growth, and an irregular aggregate may form instead. On a large scale, when ice crystals are mixed with a thick cloud with supercooled water droplets, we get *hail*. Hailstones may become very large, and weights up to 1 kg have been reported. Compare that with the vision in Revelation **16**:21 — 50 kg!

Aging of snow. The snow on the ground gradually changes its appearance, depending on the temperature, pressure and mechanical treatment (for example when one is shoveling). The points of the dendritic snow crystals become more round and the snow more compact. Such changes occur already at temperatures far below 0°C. Small snow crystals are less stable than larger ones. In a mixture of crystals of different sizes, the large crystals will grow at the expense of the smaller ones. In glaciers this process may occur over several centuries, and extremely pure and perfect single crystals weighing several kilograms have been found.

Formation of rain. The formation of ice in the atmosphere is actually also a prerequisite for the formation of rain, according to present ideas. Most clouds consist of small droplets of water, and researchers have had difficulties in understanding how the droplets can unite into so large drops that they fall by their own weight. The individual water droplets have normally the same charge and repel one another. Gradually, it was realized that rain normally does not form until crystallization occurs: when some supercooled droplets freeze to ice, they continue to grow at the expense of other droplets. Finally, the large ice crystals begin to fall toward the earth, and if the temperature is sufficiently high they melt and *rain* falls. At our altitudes this is considered to be the most common mechanism for the formation of rain.

Snow cannons. *Snow cannon* can refer to two different things: (1) A device to make artificial snow in ski slopes (also called *snow gun*), or (2) The effect of heavy snow fall as cold air blows over a warm lake towards land.

(1) *Snow gun.* Snow making started to be used on a commercial scale in the early 1970s. There are in principle two types of snow cannons. One type uses high pressure water and a powerful fan. The second type uses both water and air under

high pressure. A nozzle in the snow cannon forms droplets of water, so small that they remain suspended for a while so that they can grow and freeze. The ideal weather conditions for snow making are −5° to −15°C, less than 80% humidity and slight breeze. It is nowadays possible to make snow up to 0°C but the snow is then of slightly lower quality. A low humidity is important as a drop of water in dry environment partly evaporates and is then cooled down by the heat of vaporization. 10 cm of cannon snow corresponds to about 80 cm of natural snow. The freezing of the water is helped if you mix a nucleator of some kind into the water supply. The water may already contain some stuff that can act as nucleators, but many resorts also add special organic or inorganic materials as nucleating agents. The water is sometimes mixed with *ina* (ice nucleation-active) proteins from the bacterium Pseudomonas syringae. These proteins serve as effective nuclei to initiate the formation of ice crystals at relatively high temperatures.

Artificial snow is different from the natural snow in that it consists of small ice crystals and does not form flakes as natural snow. Artificial snow is also significantly harder in texture which can cause the adverse effect that it becomes easier "icy." Therefore this type of snow is also called *snice* (*snow-ice*). One advantage of the hard consistency is that it enables high speed skating, as it provides good grip for well-sharpened carving skis. Another advantage of the hard consistency is the durability; it melts much more slowly than natural snow which prolongs the season for ski resorts. As the properties are not the same as natural snow it may have implications for how one should wax the skies.

The strength of snice is almost like cement and snice is therefore used as building material, as e.g. when constructing ice hotels. The mixture used in the ice hotel in Jukkasjärvi in northern Sweden contains a larger amount of water than in normal artificial snow. The ice blocks are taken from the Torne river as this ice is especially clear (Fig. 3.10).

Fig. 3.10. From the ice hotel in Jukkasjärvi.

Fig. 3.11.   A train stuck in Minnesota in 1881.

Fig. 3.12.   The snow cannon in Gävle in 1998.

(2) *Lake effect snow.*   When cold air blows over a warm lake toward land, a heavy snowfall can be a result. Such "snow cannons" (blizzards) are common in North America. When very cold air from Canada flows down over the Great Lakes, a huge amount of snow falls over especially the eastern shores of Lake Erie and Lake Ontario. The winter of 1880–1881 is considered the most severe ever known in the United States (see *The Long Winter* by Laura Ingalls Wilder). Figure 3.11 illustrates the situation on March 29, 1881 in Minnesota. In Sweden a notorious snowfall (although on a much lower scale) affected the Gävle area in early December 1998, when in three days 130 cm of snow fell, paralyzing all transport links in the area (Fig. 3.12).

4
Snow and Ice Crystals in Nature

Frozen water can appear in a seemingly infinite number of different shapes. A collection of the beautiful macro photos taken by Alexey Kljatov (Figs. 4.1a–f), Andrei Osokin (Figs. 4.2a–f) and Don Komarechka (Figs. 4.3a–f) is shown below.

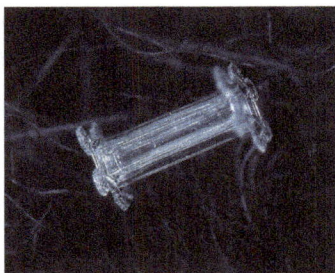

Fig. 4.1a

Fig. 4.1b

Fig. 4.1c

Fig. 4.1d

Fig. 4.1e

Fig. 4.1f

Fig. 4.2a

Fig. 4.2b

Fig. 4.2c

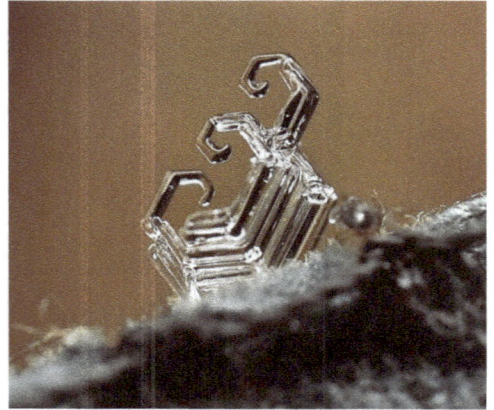

Fig. 4.2d

Fig. 4.2e

Fig. 4.2f

Fig. 4.3a

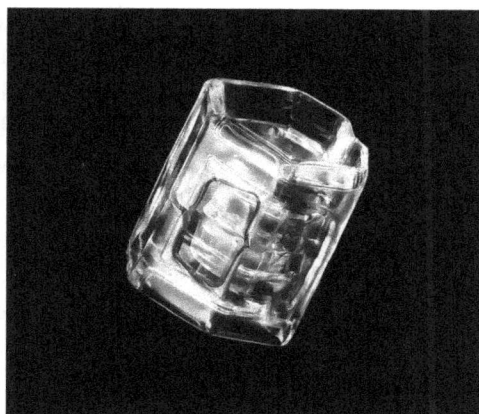

Fig. 4.3b

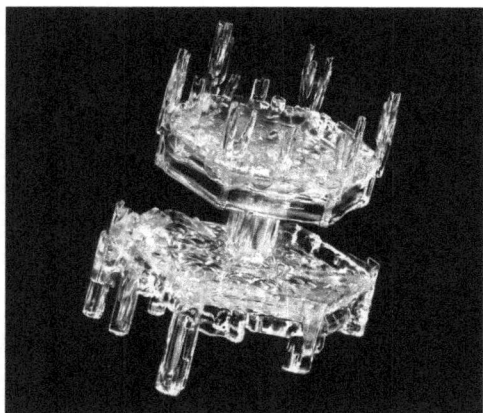

Fig. 4.3c

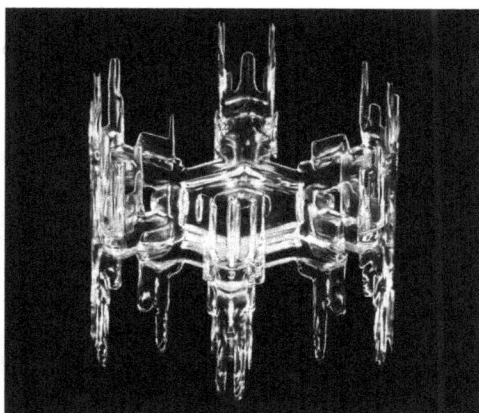

Fig. 4.3d

Fig. 4.3e

Fig. 4.3f

Thin film interference in ice crystals.   As an ice crystal grows, the edges and corners grow fastest as the addition of water molecules from the environment and the disposal of the heat of crystallization is most effective at these places. Small air pockets may then be left in the central parts of the crystal. In this way a very thin ice film may be formed between the outer surface and the air bubble. If light strikes the crystal, it is either transmitted or reflected at the outer surface. Light that is transmitted may again be reflected at the boundary between the thin ice layer and the air bubble (or at the bottom of the air bubble). The two reflected light beams will then interfere with each other. If the phase of the two reflected beams is the same, they will reinforce each other (constructive interference). If the phase is opposite, they will weaken each other (destructive interference). The phase difference depends on the thickness of the thin ice layer, the refractive index of the ice, and the angle of incidence of the light wave relative to the ice surface. Additionally, a phase shift of 180° will be introduced upon reflection at the boundary between air and ice. The pattern of reflected light which results from this interference will appear as colorful bands. If the blue light is quenched, yellow or reddish bands will appear.

Don Komarechka has taken a series of very nice macro photos showing very clearly this thin film interference (Figs. 4.4a–g). The pictures illustrate how the color changes with the thickness of the thin ice layer. A particularly interesting case is shown in Fig. 4.4g, where the crystal contains very small air bubbles distributed at different distances from the center of the crystal. As the angle of incidence relative to the air bubbles varies with the distance from the center, the color of the bubbles will also vary with this distance, as clearly shown in the picture.

Fig. 4.4a

Fig. 4.4b

Fig. 4.4c

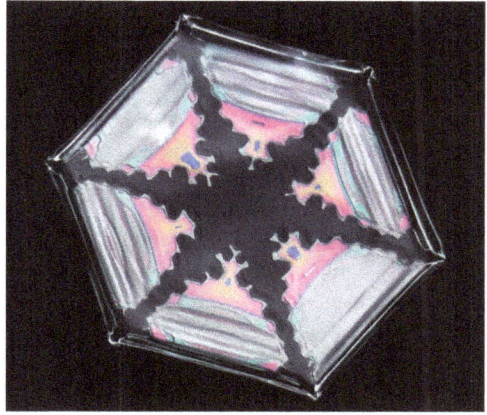

Fig. 4.4d

Fig. 4.4e

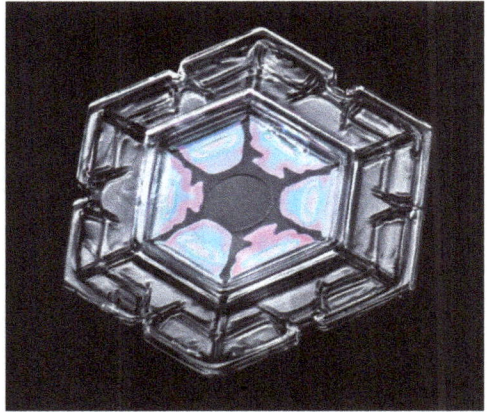

Fig. 4.4f

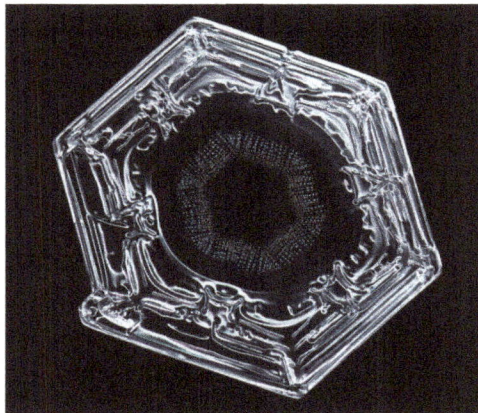

Fig. 4.4g

5
Snow for Pleasure and Art

Not only children enjoy making snow sculptures during the winter. Each year snow festivals are held all over the world. Two of the largest are in Sapporo in northern Japan and Harbin in northeastern China. Three examples from Sapporo are shown below. Even the Japanese army has in some cases been employed to build the largest monuments (Figs. 5.1a–c).

「首里城正殿」 (大通西8丁目) 制作　陸上自衛隊第18普通科連隊第2中隊
"Shuri Castle" at-Odori.

Fig. 5.1a

「ルパン三世」 (大通西10丁目) 制作　札幌市消防局
"Rupan III" at Odori.

Fig. 5.1b

Fig. 5.1c

The temperature in Harbin stays below 0°C for several months during the winter (it sometimes drops to −40°C), and the fantastic constructions may then be enjoyed for a long time (Figs. 5.2a–b).

Fig. 5.2a

Fig. 5.2b

The British artist Simon Beck has created fantastic patterns in snowfields at Les Arcs in southeastern France and at Lake Louise in British Columbia. A very skilled orienteer, he forms the patterns by walking in snowshoes, using a sighting compass (as used by orienteering mapmakers) and a sketch of the pattern on a piece of paper. A typical pattern takes around 10 hours to produce. A selection of his pictures is reproduced (Figs. 5.3a–d). There is a book on sale specializing in his work; see http://snowart.gallery.

Fig. 5.3a

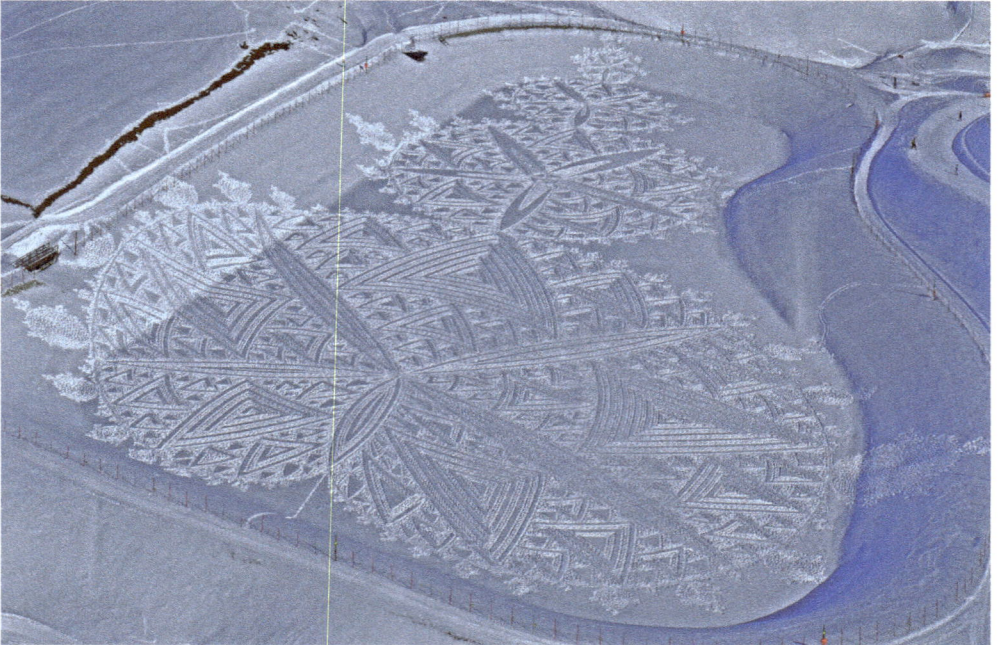

Fig. 5.3b

Fig. 5.3c

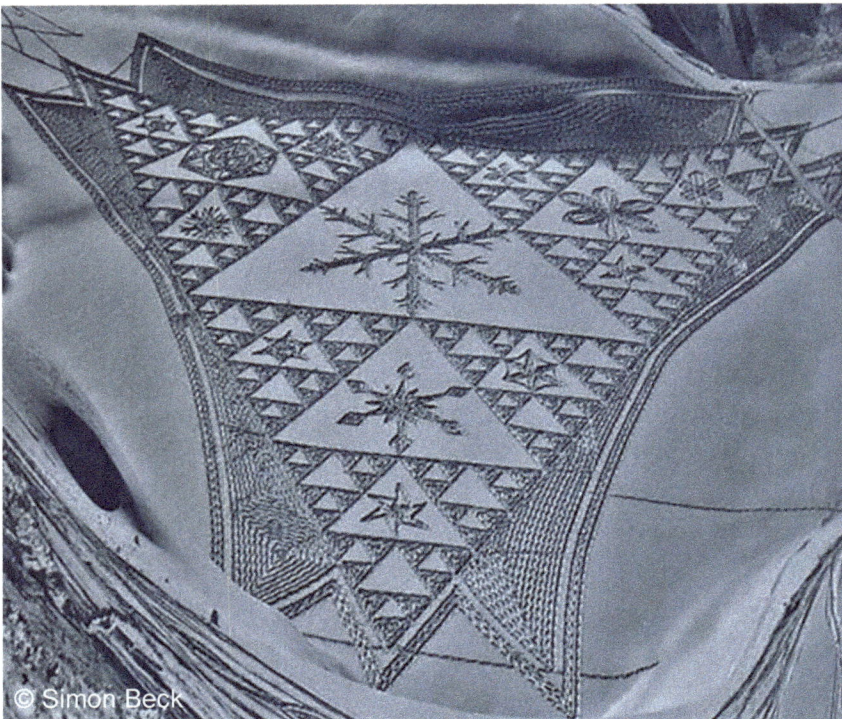

Fig. 5.3d

6

The Ice Surface and Formation of Ice Spikes

Ice spikes. A few years ago, when I was skating on a lake close to Uppsala, I noticed a lot of spikes sticking up everywhere (Fig. 6.1). Reports on the formation of ice spikes were already found in scientific journals in the early 20th century — among others, by O. Bally in 1935. Ice spikes are sometimes formed when a bowl of water is standing outside in cold weather. In recent years a large number of photographs of natural ice spikes have appeared on the Internet, and a few years ago there was a lively discussion on how to explain the formation. Briefly, this is what happens.

The water first freezes around the edges on the top surface of the bowl, until a small hole is left unfrozen. At the same time, ice starts to form around the sides, inside the bowl. Since ice expands as it freezes (as the ice takes up a 9% larger volume), water is pushed up through the hole. The outgoing water freezes around the rim, forming an ice spike. The spike can continue to grow until all the water freezes. As the ice spike is formed when water is pushed up through a hole in the growing ice, the limited space in the bowl helps to build up the pressure needed.

Fig. 6.1. Ice spikes on a lake in the neighborhood of Uppsala.

Accordingly, ice spikes are most easily formed in small containers. Small ice spikes can occasionally also be found on ice cubes in refrigerators, using distilled water in plastic ice cube trays. The ice spikes are mostly thin and cylindrical, but triangular and inverted pyramids have also been reported. The angle between the spike and the surface varies, and does not seem to have any direct relation to the internal crystal structure.

It is somewhat surprising that so many ice spikes were formed all over the lake as shown in Fig. 6.1. A large number of small pinholes evidently remained as ice formed on the surface and water pressure was somehow built up below (probably owing to the thickening of the ice layer).

The spikes normally become only around 5–6 cm long but occasionally much longer spikes have been reported. The following article, "Due North," by Harold L. Kirk, appeared in *Harborcreek Historical Society Newsletter*, April/May 2007, p. 6 (reprinted with permission):

> At 8 a.m. on a cold Saturday morning in March of 1963, Gene Heuser left his warm home on East Lake Road and headed due north. When he reached the shore of Lake Erie, the only thing ahead of him was an icy barren wasteland as far as the eye could see. His plan was to hike over the ice, about 32 miles, to the Canadian lighthouse at Long Point, stay the night and then hike back to Harborcreek. Little did he envision the many obstacles that lay ahead. Using only a small compass to guide him, he soon found that heading due north was seldom possible as he encountered pillars of ice five feet high and snow drifts of over 10 feet. He told a reporter later, "I never expected to see what I saw. It was not just a smooth surface."
>
> As evening approached and he was still miles from land, he knew he would be spending a long cold night on the ice! The moon shone brightly for about an hour but later, clouds covered the sky leaving him in near total darkness. Using a small flashlight allowed him to continue his northward trek. He said, "The flashlight lit up these huge ice chunks with a fluorescent glow into eerie forms and shapes like those of a barren planet. Sometimes I fell on the jagged surface and just lay there on the ice. I knew I could not lie down long or I would freeze." He also said that one of the most vivid recollections of that long night was of the small pinholes in the ice through which the water below was periodically forced under pressure to spout up into the air and freeze. The frozen spurts looked to him like telephone poles standing straight up all over the lake. He told the *Erie Morning News* later, "I knew my planned route from Shade's Beach to Long Point was about 32 miles but I figured I must have walked over 50 miles because of the drifts and ice chunks I had to walk around."
>
> Well past daybreak on Sunday morning, Gene reached the lighthouse where he saw some shacks belonging to a team of Canadian scientists making a lake study. They didn't believe that he had just strolled over from the nation to the south until he showed them his identification. Canadian

police escorted him to the mainland at Port Rowan, Ontario. Gene quickly revised his original plan of walking back to Harborcreek and instead called his brother in Buffalo to pick him up.

The ice surface.   Notice that the ice surface in Fig. 6.1 is quite uneven. This is a common feature of almost all ice surfaces.

When water in a closed environment, such as a small container, begins to freeze, separate ice crystals will form at the edges and the growing ice crystals will soon collide with one another. At the same time the water below continues to freeze and pressure is built up from below and the ice on the surface cannot form a completely flat surface, as illustrated in Fig. 6.2 (this may be considered the first stage of spike formation, in which case water is pressed through small pinholes still remaining in the surface). For this reason, the ice surface on a small pond in the street or even on an inland lake will not be completely flat throughout. The best condition for finding an extended completely flat ice surface is consequently at the seashore, where the ice may grow outward without colliding with other growing crystals.

Fig. 6.2. Typical pattern of the ice surface formed in a container.

Icebergs.   Icebergs form when chunks of ice "calve" or break off from glaciers. They come in all shapes and sizes, from small chunks to the size of a small country. Only 20% of a floating iceberg is visible above water, owing to the difference in density of ice and water. Most icebergs are found in the North Atlantic and the cold waters surrounding Antarctica, and pose a danger to ships — the sinking of the *Titanic* in 1912 is still fresh in the memory. But there have also been plans to utilize them for practical purposes. The possibility of towing icebergs to arid regions with a lack of fresh water has been considered for several decades, and such transportation has been tried in a few cases. For example, between 1890 and 1900, small icebergs were towed by ship from Laguna San Rafael, Chile, to Valparaiso and even to Peru, a distance of 3,900 km. Although such transportation is technically possible, the economy is doubtful. One major problem is that a large part of the iceberg melts during the transportation.

Effect of ice melting on the sea level.   With the present increase in the global temperature, which may lead to rapid melting of ice in Arctic regions, there has been great concern that it will result in a large increase of the sea level. Here one should perhaps be reminded that only the melting of land-based ice will affect the sea level. The melting of sea ice and floating icebergs will not change the sea level.

Ice as aircraft carrier and Project Habakkuk.   During the Second World War, a top secret project was launched to build ice-floes to be used as landing strip, for aircraft. Pure ice was evidently too brittle and too quick-melting for the purpose. However, the physical chemist Herman Mark found that a mixture of ice and wood pulp would make a very hard material, which was called "pykrete." It could be machined like wood, and when immersed in water it formed an insulating shell of wet wood pulp on its surface that protected its interior from further melting. Winston Churchill approved the project, named Habakkuk — after one of the books in the Hebrew Bible ("Behold ye among the heathen, and regard and wonder marvelously: for I will work a work in your days, which ye will not believe, though it be told to you."). Two crystallographers, J. Bernal and M. Perutz, were involved as scientific advisers in the project. A large amount of information about this fantastic project is available on the Internet.

7

Structure and Physical and Chemical Properties of Water and Ice

The water molecule.   As oxygen is more electronegative than hydrogen, the electrons are pulled slightly toward oxygen, and hydrogen gets a small net positive charge. The oxygen gets a corresponding negative charge (Fig. 7.1). This results in polar properties of the water molecule (it has a dipole moment of 1.85 debye). Many of the special features of the water molecule are due to this polarity. Of particular importance is the distinctive tendency of the water molecules to form hydrogen bonds with one another or with other polar molecules, and they can then operate both as donors and acceptors. Sometimes the two lone pairs in the water molecule are illustrated as "rabbit ears" standing out tetrahedrally relative to the O–H bond directions (Fig. 7.2).

However, the actual electron distribution does not at all agree with this model. There is a fairly even distribution over the entire area. The electron distribution of the free water molecule from quantum-mechanical calculations is shown in Fig. 7.3: (a) in the plane; (b) perpendicular to the plane of the molecule. The corresponding deformation density is shown in Fig. 7.4 (the deformation density shows the deformation of the electron clouds of the free oxygen and hydrogen

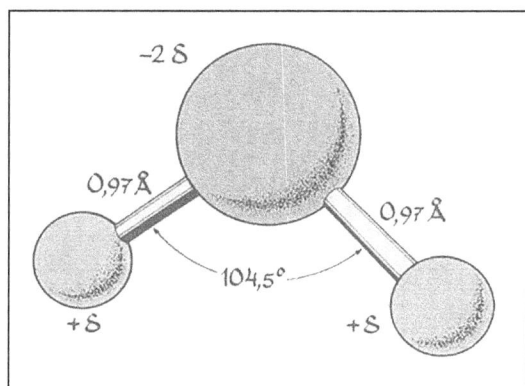

Fig. 7.1. Geometry of the free water molecule.

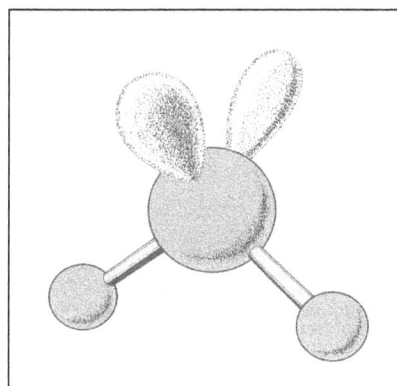

Fig. 7.2. "Rabbit ear" model.

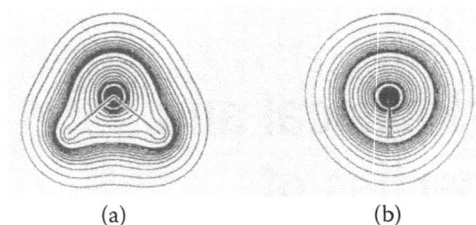

(a)   (b)

Fig. 7.3. Total electron distribution.

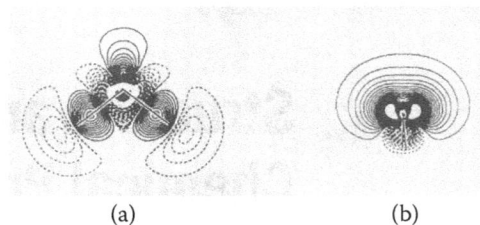

(a)   (b)

Fig. 7.4. Deformation density.

atoms when the molecule is formed — the difference between the actual density minus the "promolecule" density).

As the simple picture of the water molecule with "rabbit ears" does not illustrate the actual electron distribution, this model should be avoided. With the rabbit ear model one might suggest that the water molecule binds specifically in the free electron pair directions (tetrahedrally relative to the OH directions) owing to an electron concentration in these directions. It is true that the water molecule often forms a tetrahedral arrangement with other molecules, but this can be explained by purely topological reasons: Suppose that we want to build a three-dimensional arrangement of water molecules and that we require that all these have the same environment. If each water molecule functions as donor (O–H$\cdots$) in two hydrogen bonds (O–H$\cdots$O), then each water molecule also must accept two hydrogen bonds. For purely geometrical reasons, these two acceptors will be arranged approximately tetrahedrally relative to the OH directions in order to form the most favorable arrangement.

For a more exhaustive discussion on the role of the lone pairs in hydrogen bonds in general, see the enclosed reprint.

The structure of ordinary, hexagonal ice, I_h. In all forms of ice, each water molecule is tetrahedrally surrounded by four other water molecules (Fig. 7.5). Owing to the hydrogen bonds, the O–H bond is slightly longer in ice than in the free water molecule. The only structure that is stable at ordinary pressure and at moderately low temperature is hexagonal ice, I_h,

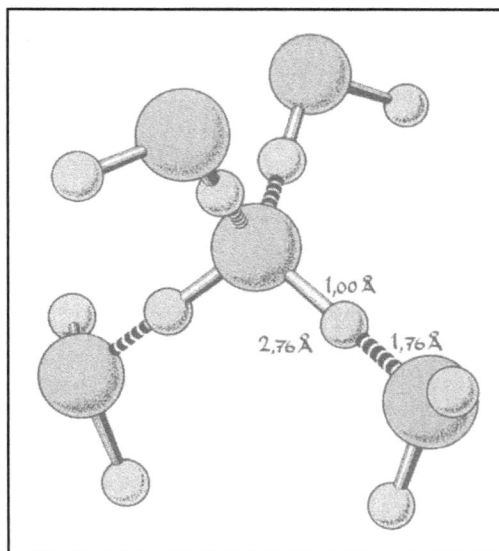

Fig. 7.5. Environment of H_2O in ice.

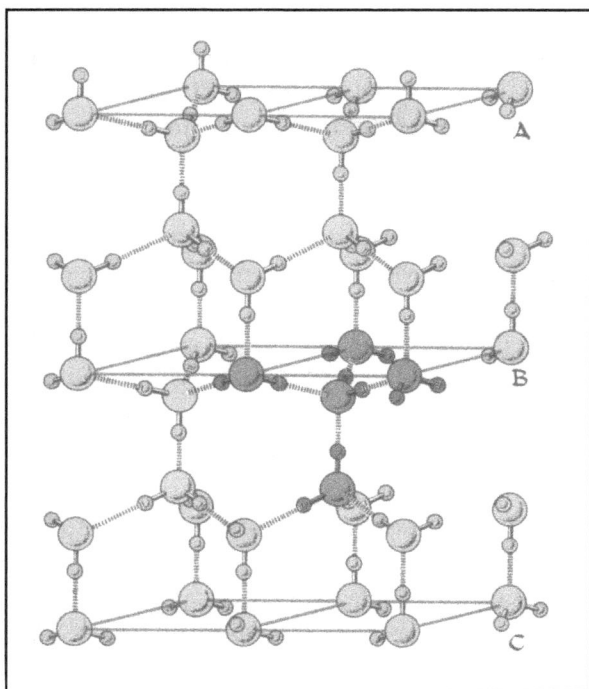

Fig. 7.6. Structure of ordinary ice.

which has sixfold internal symmetry. The crystal structure is shown in Fig. 7.6. There is one very important feature of this structure: the water molecules are disordered! The picture shows an example of how the molecules can be oriented differently in different unit cells. Compare the molecules marked A, B and C. If all possible orientations are equally frequent in the structure, an arrangement O–H···O and O···H–O will be equally probable in a particular hydrogen bond (if all unit cells are taken together). A diffraction study will then show one-half hydrogen atom at each place. This disorder is maintained all the way down to 0 K, and this results in a so-called residual entropy (see more details below).

Ordinary ice has in many ways unusual properties compared to similar substances containing light molecules. As ice has so many interesting physical properties, it was natural that many scientists tried to determine its crystal structure very early on. But that proved more difficult than expected. Already in the early 1920s, Bragg and other crystallographers showed that ordinary ice has a hexagonal symmetry (D6h, P63/mmc), and the positions of the oxygen atoms were determined. The structure forms a very open network, due to the hydrogen bonds. If the molecules are instead tightly packed next to one another, each molecule will be surrounded by no less than 12 neighbors. Such a packed arrangement occurs in the crystals of the analogous compounds H_2S, H_2Se and H_2Te. In these cases no strong hydrogen bonds are formed and the molecules try to pack as closely as possible.

There was general agreement about the positions of the oxygen atoms, but the placement of the hydrogen atoms caused a lively debate. The problem was that at that time only X-ray diffraction was available and as the hydrogen atoms spread X-rays rather weakly they could not find the hydrogen atoms. Today the X-ray method is much more developed and the location of hydrogen atoms can be determined in not-too-complicated compounds, though still with moderate accuracy. Many more-or-less fanciful models were proposed. Some researchers suggested that the water molecules do not have the same orientation in every unit cell, which is not possible in an ideal crystal structure. Thermodynamic measurements showed that when ice is cooled to near 0 K there remains a certain entropy, 3:41 J/degree·mol, a so-called residual entropy. Entropy is a measure of the degree of disorder in a system: in a completely ordered system, the entropy must approach zero at 0 K. The experimental measurements thus indicated a disordered structure. Pauling suggested that water molecules could have a statistically equal distribution among all possible orientations of the water molecules, while still forming hydrogen bonds to the four neighbors (see Fig. 7.6). Based on this model he calculated an entropy of 3:38 J/degree·mol, which is close to the experimental value. The dispute about the correct ice structure was long and lively. The definitive breakthrough did not come until we had access to neutron diffraction.

At the end of the 1940s, one of the first research reactors in the world was built in Oak Ridge, USA, and here pioneering work was done in neutron diffraction research. E. O. Wollan, W. L. Davidson and C. G. Shull made in 1949 a study of ice in powder form. In 1994 Shull received together with Bertram Brockhouse the Nobel Prize in Physics for this neutron study of ice, among other work. Even more detailed information was obtained some years later when S. W. Peterson and H. A. Levy in Oak Ridge studied single crystals of ice. In both of these early studies heavy water, D_2O, was used as the deuterium isotope gives only a small contribution of disturbing (incoherent) scattering compared to ordinary hydrogen. Peterson and Levy are probably the first to investigate a single crystal by neutron diffraction.

The neutron results were awaited with great excitement — the question of the ice structure had now become a classic problem. The investigations confirmed Pauling's theory of the disordered structure of ice. A diffraction study does not show the content in a specific unit cell, but the mean value of the contents in all unit cells. In a completely disordered structure the probability is only 50% that hydrogen is in one of the two possible positions in a particular oxygen–oxygen bond. In the neutron study it was indeed found that the hydrogen positions were on average only half-occupied. Nevertheless, there are still many questions about the details of ice structure to explore.

Thermodynamic measurements of $Na_2SO4 \cdot 10H_2O$ have shown that this compound also retains a certain residual entropy when cooled to low temperatures. During my time in Berkeley, 1957–1959,

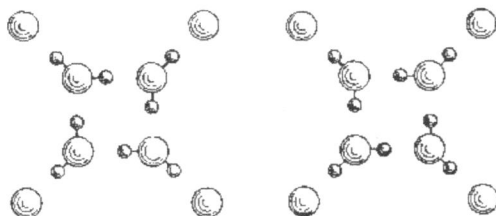

Fig. 7.7. Alternate configurations for water molecules.

the crystal structure of this compound was determined by X-ray diffraction. In the structure there are two kinds of rings, each with four hydrogen bonds. There are two possible configurations of the hydrogen atoms in these rings (Fig. 7.7) and complete disorder of these configurations corresponds to the residual entropy found experimentally ($R \ln2$ per mole).

The structure of other forms of ice. Depending on temperature and pressure, water can assume about 17 different crystal structures — perhaps more than any other known material. The phase diagram is shown in Fig. 7.8.

Amorphous ice and ice I_c. If water vapor condenses on a cold surface at temperatures below about −160°C, a non-crystalline, amorphous form — glass — is created. If the condensation occurs between about −160°C and −120°C, a crystalline, cubic form — ice I_c, with a diamond structure is created, but otherwise it is similar to ice I_h. Cubic ice possibly also occurs at high altitude in the atmosphere.

Ice II–XI. These structures are formed at higher pressures — between 0.2 GPa and 2.2 GPa — except for ice X, which is first formed at pressures above 44 GPa. The water molecules in ice II, VIII, IX and XI are more-or-less ordered, whereas the other crystal structures are partially or completely disordered. The coordination is in all cases tetrahedral, as in hexagonal ice, but the relative positions of the

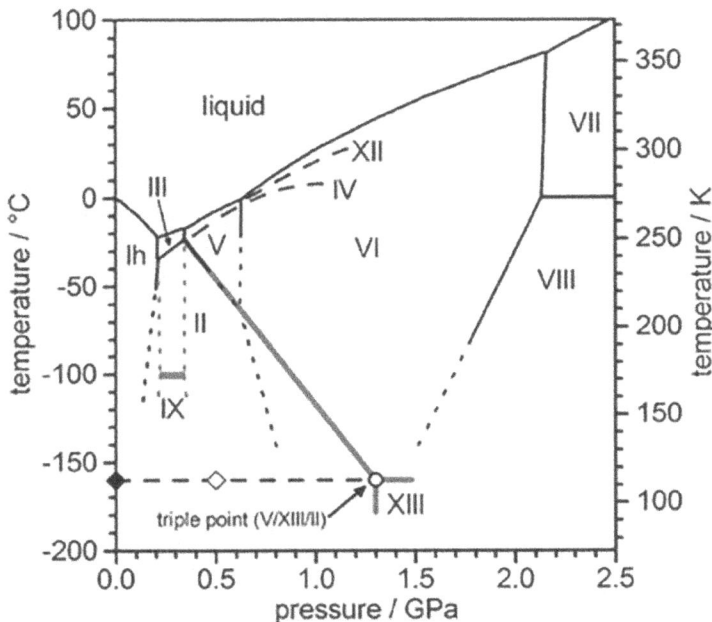

Fig. 7.8. Phase diagram of ice.

tetrahedra are different and the water molecules are more densely packed in the high-pressure forms. The structure of ice X is not yet known but is postulated to be of an entirely different character with symmetrical hydrogen bonds.

Doping ice I_h with 0.001–0.1 moles of KOH has produced crystals with an orderly proton distribution corresponding to the structure of ice XI.

8
Physical Properties of Water and Ice; Significance in Nature

The melting and boiling points of water are unusually high owing to the hydrogen bonds. Without hydrogen bonds the melting point should be around −100°C and the boiling point perhaps around −80°C, when compared with the analogous compounds H_2S, H_2Se and H_2Te, where the bonds between the molecules are much weaker (van der Waals bonds), and if the molecular weight of the water molecule is taken into consideration. To melt 1 kg of ice, 334 kJ is needed and the corresponding amount of heat is released when water freezes.

An interesting phenomenon occurs in the fall when the water in a lake starts to freeze *and there is no wind at all*. When some water turns into ice, the heat released is transferred to other water molecules, which may then turn into gas form. As the air is cold, this gas will condense to water droplets and mist is formed — "the lake smokes." According to the old tradition, you can then tell that the lake has started to be covered with ice. It should perhaps be added that "smoke" may of course also be formed when cold wind blows over a lake with warm water — and perhaps cause a snow cannon.

The water is very easily supercooled: when ice is formed the irregular network of hydrogen bonds in liquid water must be reorganized into a regular tetrahedral network. Clean water can thus be supercooled down to −20°C before it crystallizes, and water droplets (1–10 microns) in the atmosphere even down to −40°C. The very special characteristics of the ice structure make it difficult to take up other compounds; an exception is ammonium fluoride, whose structure is similar to that of ice. When ice is formed from saline water, the majority of the salts will accordingly remain in the solution and the ice will become virtually salt-free. This can in principle be used for the production of drinking water from seawater, but this method of desalination is very energy consuming. However, if the ice formation takes place very quickly, the salts will accompany the ice crystals as an impurity, although not as a solid solution inside the ice structure. The principal current process to separate salts from water uses semipermeable membranes and pressure, applying reverse osmosis technology.

The very open structure of ice leads to a low density — 0.917 g/cm^3 — of pure ice at 0°C. When ice melts at 0°C, part of the bonds in the tetrahedral network are broken and the water molecules can pack somewhat closer (but hydrogen bonds still play a major role in liquid water, and much of the short-range order persists even when the ice has melted). The density of liquid water at 0°C — 1.000 g/cm^3 — is therefore higher than for ice — a remarkable fact, as in most cases a solid has a greater density than the corresponding liquid. Ice will accordingly float on water, which is of great importance in nature, as we shall see. When water freezes to ice, the volume increases by about 9%, and in closed systems this will have strong expansion effects — "frost shattering" in nature. For the same reason, the melting point of ice is lowered when subjected to pressure (0.0074°C/bar), which is important in glaciers, for example.

When the temperature of water rises from 0°C, additional hydrogen bonds between the water molecules are broken and therefore the density increases at first. At the same time, however, another factor comes into play: when the temperature rises, the motion of the water molecules increases and the molecules will take up more space (the vast majority of solids and liquids expand with temperature for the same reason). Eventually, the latter effect will dominate, and water expands with temperature in the same manner as a normal liquid. The combined effect of these two factors has the result that pure water gets a maximum density at +4°C — a most remarkable phenomenon that is of great importance in nature. With increasing salinity the freezing point decreases as well as the temperature of the maximum density, so that water with a salinity of 2.5% on cooling constantly increases its density down to the freezing point (for the oceans with a salinity of 3.5%, the freezing point is −2°C).

When the water in a lake gradually cools down in the fall, the following events will occur owing to the special properties of ice and water (if we disregard the effects due to wind). When the surface water is cooled, it will sink toward the bottom, until the bottom water has a temperature of +4°C. Upon further cooling from +4°C to 0°C, the water at the surface will remain there as it has a lower density than the bottom water. Eventually, ice will be formed, and since it does not sink to the bottom the ice will form an insulating layer that delays the continued cooling of the water. The growth of the ice layer occurs all the time on the underside of the ice. Ground freezing is prevented effectively and normally occurs only in shallow water (the effects of currents can of course accelerate this process). It is quite remarkable how this collaboration of several unusual physical properties of ice and water affects our environment and helps aquatic animals to survive the winter.

While the serum of fish living in polar seawater can carry enough salt to lower their freezing temperature by about 1°C, this is not enough to prevent freezing. Consequently, they must rely on another mechanism for survival in a supercooled state. The antifreeze effect is a crucial issue and is related to crystal growth controlled by biological macromolecules.

Surface properties of ice and snow.   Why do we glide more easily on ice and snow than on asphalt or gravel? A common notion has been that when the skate glides across the ice, the pressure causes melting of the ice and provides a liquid film with very low friction. But this explanation is not entirely correct, according to current research. When we go skiing or skating, the pressure is too low to melt the ice. Another theory been suggested regarding the structure of the outermost layer: the water molecules on the surface do not have a complete system of hydrogen bonds and the incomplete bonding situation could perhaps allow the molecules on the surface to rotate. Individual molecules or small groups of molecules could then function as a sort of ball bearings. However, it seems more likely that the molecules turn aside but are still bound to the underlying molecules.

Research on the surface properties of ice remains very active. Extensive experimental investigations have been made in recent years, e.g. by ellipsometry, X-ray scattering, LEED (low energy electron diffraction), NMR (nuclear magnetic resonance) and AFM (atomic force microscopy). Theoretical calculations — among others, the application of molecular dynamics simulation — has also given detailed information about the arrangement and dynamics of water molecules in ice and water. These results show that the water molecules of the outermost layer have great mobility and increasing disorder with increasing temperature. It is thus a question of a more-or-less disordered network of water molecules of high mobility at the surface but not a typical liquid layer where the molecules are relatively free to move; we can call it a "quasi-liquid layer." The surface properties of most materials can be quite different from the inner part of the material, "bulk properties," and approach the properties of the liquid phase. The ice surface may perhaps be likened to a brush: the hairs are stiff at low temperature and this leads to a rough surface, but the hairs bend more easily when the temperature rises and the friction is then lower when an object such as a ski glides over the "brush."

The structure of liquid water.   How are the water molecules arranged in water? They move around or twist in the liquid, on a timescale as short as a picosecond (10^{-12} s), and since the arrangement can change more or less radically, it is obviously impossible to report a specific, fixed "structure." To determine the instantaneous structure we must employ experimental methods which can register the structure in an even shorter time. X-ray and neutron diffraction (elastic scattering) only give information about the average distribution of molecules over a long time span (a superposition of all structures during this time), but on the other hand they can give definite information about the distances between the molecules (the radial distribution). Since diffraction and spectroscopy have perhaps provided the most important contributions to the models of the water structure, we will touch upon the information these methods can provide.

Radial distribution. The electron distribution in atoms is usually illustrated with the so-called radial distribution function, $4\pi r^2\psi^2$, which indicates the relative probability of finding the electron at a distance r from the center. Similarly, the distribution of molecules in a liquid is described by a radial distribution function, $g(r) = 4\pi r^2\rho(s)$, which indicates the relative probability of finding two atoms at a distance r from each other (the "pair correlation"). Maxima occur where the probability is very high. However, the diffraction investigations cannot give direct information on which pair of atoms is at this particular distance.

X-ray diffraction. X-rays are scattered by electrons, and therefore heavier atoms scatter the radiation more strongly than light atoms. The contribution to $g(r)$ is proportional to the product of the scattering power of the two atoms in the atom pair. In the case of water, oxygen scatters the X-rays about eight times more strongly than hydrogen with only one electron. This means that the contribution to $g(r)$ is completely dominated by the oxygen–oxygen pair. The relative probability of finding neigboring water molecules at a certain oxygen–oxygen distance at different temperatures is shown in Fig. 8.1. The large peak at 2.8 Å corresponds to the O–O distance between the closest hydrogen-bonded water molecules. From the location of the smaller peaks at larger distances, one tries to draw a conclusion about the arrangement of water molecules further out and to derive a model for the water structure.

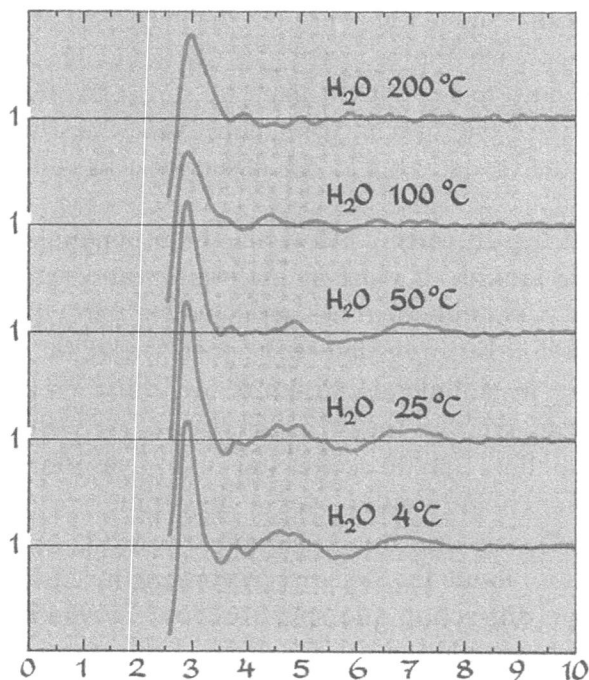

Fig. 8.1. Radial distribution of O–O distances (Å).

The distance of 2.84 Å in liquid water is slightly longer than the distance in ice, 2.76 Å. These values agree well if we assume that the regular tetrahedral network in ice is partly broken down in liquid water so that each water molecule on the average is surrounded by slightly more neighbors than in ice (approximately 4.5 compared to 4 in ice, and each distance then becomes somewhat longer).

Neutron diffraction.   Neutrons are scattered by the nuclei, and light nuclei may scatter about as strongly as heavier nuclei. This means that the hydrogen–hydrogen contribution to $g(r)$ will be approximately as large as the oxygen–oxygen contribution. The radial distribution function will thus reproduce all distances, O–O, O–H and H–H, with approximately equal weight. However, with all these three types of distances simultaneously included, the $g(r)$ plot becomes extremely complicated and practically impossible to interpret. But, through replacement of normal hydrogen (H) by deuterium (D), one can vary the relative contributions of $g(r)$. The scattering power of hydrogen and deuterium is very different, and by diffraction studies of water with three or more different isotopic compositions it is then possible to separate $g(r)$ for the three different types of distances. In recent years, neutrons have mainly been utilized in the diffraction investigations.

Models for the water structure.   Many different models have been suggested for how water molecules are arranged in liquid water, and there is still a lively debate. Many of the earlier models only tried to explain specific anomalous properties of water, such as the high heat capacity and compressibility, or the increase in density when the ice melts. But it has been difficult to find a model that can explain all the experimental facts in a reasonably satisfactory manner. It is particularly important to find a model that is consistent with the current "structure-specific" data, where the results from diffraction and spectroscopy are vital. Essentially two main types of models were suggested early on: "homogeneous models" (or "uniform continuum models") and "cluster models" (or "mixture models"). These models are briefly described below.

Homogeneous models.   The original model was formulated by J. D. Bernal and R. H. Fowler in a classic work published in 1933 and by J. Lennard-Jones and J. A. Pople in 1951. Here it is assumed that all water molecules are all the time hydrogen-bonded to one another in a network without regular repetition, and where the molecules retain a fourfold coordination even when the details of the arrangement change with time by the bending and stretching of the hydrogen bonds. It is thus assumed that none of the four hydrogen bonds around a certain water molecule is broken when the molecules move, and that the bending and stretching of the bonds is conducted independently in the different water molecules.

Cluster models. Here it is suggested that clusters of three- or four-coordinated water molecules are bound together in a rigid network with a lifetime of about 1 nanosecond (10^{-9} s). It is further assumed that these clusters are separated by borders in which the water molecules are involved in only one hydrogen bond or perhaps even none at all. In the "flickering cluster" model by H. S. Franks, the boundaries and the arrangement of the clusters change over time by cooperative movements of the water molecules. In contrast to the homogeneous models, in cluster models it is assumed that hydrogen bonds are not so easily bent and that linear hydrogen bonds have a significantly lower energy than bent ones.

An alternative model, which can best be attributed to the cluster models, is the "interstitial model." Here it is assumed that the individual water molecules can be shaken loose and enter cavities in the very open ice structure. Such a model was proposed, among others, to explain the increase in density when the ice melts.

Theoretical calculations. With the advent of fast computers, it has become possible to make advanced theoretical calculations of the water structure. In so-called molecular dynamics simulations, a starting structure is first selected (such as the arrangement of ordinary ice) and subsequently the individual water molecules are allowed to move in accordance with Newton's equations of motion, and thus in a purely classical physical manner. However, the problem is to find a good model for the forces between the water molecules. In the calculations it is important, among other things, to take into account "many-body interactions" ("cooperativity"): when two water molecules are bonded to each other, the electron distribution in these molecules is affected and this alters the forces to other water molecules. The theoretically calculated radial distribution from molecular dynamics simulation is in good qualitative agreement with the experimentally determined distribution.

Present situation. Neither of the models described above seems to be separately sufficient to explain all the anomalous properties of liquid water. Against a strictly uniform model speaks, among other things, the fact that the water molecules most likely have an environment that varies strongly with time owing to rotation of the water molecules and breaking of the hydrogen bonds. Against a rigid cluster model with close-to-linear hydrogen bonds speak surveys of crystalline hydrates which show that the hydrogen bonds in most cases are more or less bent. In the cluster models it has to be assumed that the clusters are constantly changing rapidly and the "unbound" water molecules between the clusters can reasonably exist only for a very short time, of the order of one-tenth of a picosecond, for example in connection with reorientation. On a slightly longer timescale, it is very unlikely that a water molecule is not hydrogen-bonded to any neighbor at all; from structure studies of compounds containing water, it is known that water molecules have a *very strong* tendency to be involved in hydrogen bonds. From molecular dynamics simulations and experimental studies, we also know that the water molecules have on average an environment reminiscent of the tetrahedral arrangement of ice and where the network is still relatively open.

Polywater. In this context, one should also mention the lively debate that took place between 1966 and 1971 concerning the possible existence of "super water." The excitement was over the fact that in some experimental studies of water a liquid had been observed with a maximum density of 1.4 g/cm^3, and which at $-30°C$ crystallized into a previously unknown form of "ice." Some people were concerned about the risk that this super water could get out into nature and transform ordinary water with disastrous consequences. A large number of normally serious researchers threw themselves into the debate and more or less imaginative proposals for the structure of the super water were published in scientific journals. After thorough investigations it could be concluded that the liquid simply consisted of water that had been contaminated in the apparatus used (made of glass).

9
The Water Molecule is Unique

Water plays a unique role in chemistry. The special properties of the different forms of water — from ice and snow to liquid water — are due to hydrogen bonding (O–H···O) between the H_2O molecules. Of particular importance is the distinctive tendency of the water molecules to form hydrogen bonds with one another or with other polar molecules, and they can then operate both as donors and acceptors. The hydrogen bond is of fundamental importance in biological systems since all living matter has evolved from and exists in an aqueous environment. Hydrogen bonds are involved in most biological processes as little energy is needed in forming as well as breaking these bonds. An important step in this context is proton transfer between donor and acceptor. Proton transfer in hydrogen bonds is one of the simplest chemical reactions and plays an important role in many other fields of physics and chemistry. The proton transfer path in hydrogen bonds involving fluorine, oxygen and chlorine, derived from theoretical

Fig. 9.1. The lava waterfalls Hraunfossar.
(*Photo: Ingeborg Breitfeld, Reykholt.*)

potential energy surfaces, is treated in an enclosed reprint. For a general review of hydrogen bonding, see the enclosed paper by Thomas Steiner: "The Hydrogen Bond in the Solid State."

An extensive summary of recent research on water in its different forms, "Water Structure and Science" by Martin Chaplin, is available in the Internet: http://www1.lsbu.ac.uk/water.

I have tried to illustrate the fantastic world of water in all its different forms. I encourage you to spread this message to all your friends — in the same way as the rivulets of the lava waterfalls Hraunfossar in Iceland. Here the water from the glacier Langjökull flows underground around 20 km through the porous lava field Hallmundarhraun, until the rivulets are streaming into the Hvitá river, spread over a distance of almost 1,000 m (Fig. 9.1).

Reprints

Z. Phys. Chem. **220** (2006) 963–978 / **DOI** 10.1524/zpch.2006.220.7.963
© by Oldenbourg Wissenschaftsverlag, München

The Role of the Lone Pairs in Hydrogen Bonding

By Ivar Olovsson*

Materials Chemistry, Ångström Laboratory, Box 538, SE-751 21 Uppsala, Sweden

Dedicated to Prof. Dr. Dr. h.c. Wolf Weyrich on the occasion of his 65ᵗʰ birthday

(Received April 24, 2006; accepted April 25, 2006)

Hydrogen Bond Directionality / Role of the Lone Pairs / Topological Considerations / Water Dimer / Water-Hydroxide Ion

The paper discusses some aspects of the electron lone-pairs in H-bonded structures: their role in determining the short-range structure and the effect of the environment on the electron density. In the water molecule the entire non-bonded region appears to be equally accessible for hydrogen bonding and the details of the hydrogen-bond arrangement are mainly determined by simple geometrical and topological requirements. Many examples may be taken to illustrate that it is important to take the whole electron and nuclear distribution into account when discussing the relative arrangement of interacting molecules. The resulting structure of one particular compound is determined by the net balance of many intermolecular interactions and not only by the hydrogen bonding, even if the resulting structure is consistent with hydrogen-bond directionality. From structural data it can be concluded that the immediate acceptor of a hydrogen bond is some negative charge accumulation, such as in a lone-pair region, but not specifically any individual lone pairs in the traditional, atomic sense.

1. Introduction

In elementary discussions a hydrogen bond is commonly illustrated in the following way:

X-H---:Y

The immediate acceptor of the hydrogen bond is here considered to be a lone pair which is also expected to play a definite role in determining the geometrical arrangement of the donor and acceptor molecules.

---

* Corresponding author. E-mail: ivar.olovsson@mkem.uu.se

The present paper discusses the role of the lone pairs in determining the geometrical arrangement of H-bonded structures and the general features of the electron density in the lone-pair region. Due to the fundamental importance of water, not least in biological systems, the hydrogen-bond situation involving the water molecule will be used as a model system in the following discussion.

The focus of the paper lies on the following:

– Topological considerations in the formation of hydrogen bonds.
– The electron density distribution in the free water molecule.
– The electron density in the lone-pair region of the water molecule in crystals and the influence of the environment.
– Major factors determining the hydrogen-bond directionality.
– Factors determining the geometry of the water dimer.

2. Early empirical studies of hydrogen-bond directionality

A statistical analysis of experimental observations of the hydrogen position of the hydrogen bond donor relative to the acceptor ROH molecule was made by Kroon *et al.* [1] for 196 hydrogen bonds in 45 crystal structures of polyalcohols, saccharides and related ROH compounds studied by X-ray diffraction (Fig. 1).

There appeared to be no distinct preference for acceptance of the hydrogen atom just along the traditional lone-pair directions of the acceptor ROH molecule, *i.e.* in the direction forming approximately tetrahedral angles with both the OR and OH bonds. Instead, the entire non-bonded region appeared to be equally accessible for hydrogen bonding.

Many attempts have been made to determine the way small molecules bind to macromolecules and how this binding can result in specific activity. Rust and Glusker [2] investigated possible directional hydrogen bonding to sp^2- and sp^3-hybridized oxygen atoms. They made a statistical study of several hundreds of structures containing ether, ketone, epoxide, enone and ester groups, based on data in the Cambridge Crystallographic Data Base. As an example, the scatter plots in Fig. 2a illustrate the positions of the hydrogen donor atoms X (in X-H) involved in hydrogen bonding to the oxygen atoms Y in general ketones. At first sight the distribution looks approximately the same as in Fig. 1. However, the authors point out some problems in the perception ("the eye tends to give undue prominence to outliers") and devised a technique to transform the scatter plots into continuous distributions by convoluting point atomic density with a diffuse (probability) density (*cf.* Rust and Glusker for details). The contour maps corresponding to the scatter plots in Fig. 2a are shown in Fig. 2b. In this method of illustration there appears to be a certain preference to approach in the lone pair directions. An even more pronounced preference is shown in unhindered cyclic ketones (Fig. 3b), whereas the distribution is more uniformly spread out in the non-bonded region in unhindered cyclic ethers, Fig. 3a. The

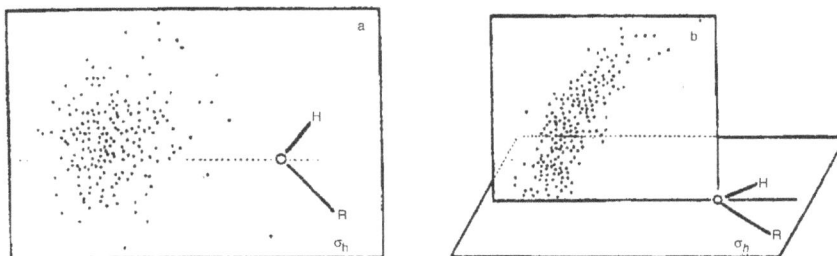

Fig. 1. Observed hydrogen positions relative to various ROH molecules (Reprinted with permission from [1], copyright 1975 Elsevier).

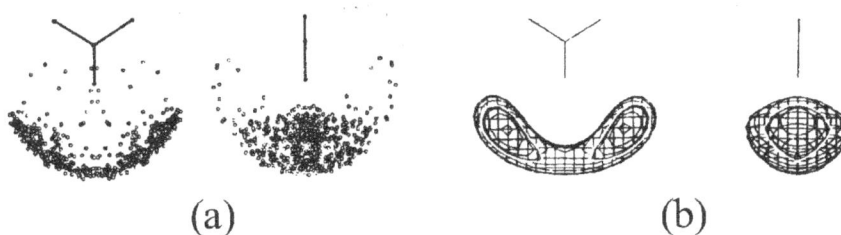

Fig. 2. (a) Scatterplots (viewed in two perpendicular directions) illustrating the positions of hydrogen donor atoms X (in H-X) involved in hydrogen bonding to oxygen Y in general ketones. (b) The corresponding contour maps (*cf.* the text). (Rust and Glusker [2])*.

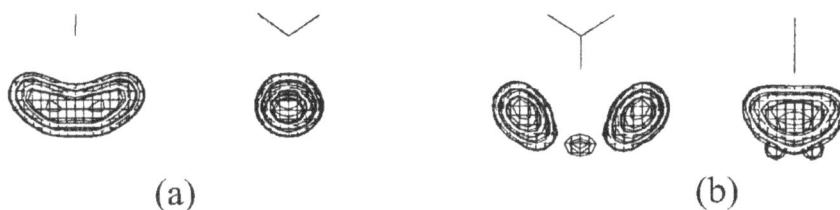

Fig. 3. Contour maps illustrating the positions of hydrogen donor atoms X (in H-X) involved in hydrogen bonding to the functional oxygen atom Y in (a) unhindered cyclic ethers, (b) unhindered cyclic ketones (Rust and Glusker [2])*.

authors concluded that hydrogen-bond directionality is indeed an important factor in many hydrogen-bonded structures. But can this be considered as a result of approach to the individual lone pairs? In an attempt to answer this we will later illustrate the electron distribution in the lone-pair region of some molecules.

---

* Reprinted with permission from [2]. Copyright 1984 American Chemical Society.

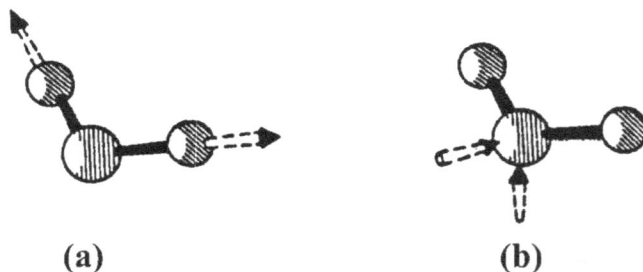

(a)   **(b)**

Fig. 4. Geometrical requirements in hydrogen bond formation: (a) *n* hydrogen bonds donated, (b) *n* hydrogen bonds accepted.

3. General topological considerations

From empirical data it is found that if a molecule contains "active" hydrogen atoms, *i.e.* contains groups normally forming hydrogen bonds, such as −OH and −NH, then all such hydrogen atoms have a strong tendency to participate in hydrogen bonding. It is very seldom found, for instance, that the hydrogen atoms in a water molecule are not both engaged in hydrogen bonding. This fact has an important bearing on the arrangement of hydrogen bonds. This will here be illustrated for the case of molecule AH_n containing n active hydrogen atoms. Suppose that we try to build up a three-dimensional structure containing only AH_n molecules and assume that all these molecules have equivalent surroundings. Thus, if each molecule acts as a donor for n hydrogen bonds, each molecule must also act as an acceptor of the same number of hydrogen bonds, as shown in Fig. 4.

We would thus need n lone pairs on each AH_n molecule. However, one seldom finds molecules with the same number of hydrogen atoms and lone pairs. Water is a rather unique molecule in this respect. In most other compounds the ideal situation is not found; ammonia is a typical example. Here only one lone pair is available and it might then be concluded that only one of the three hydrogen atoms will be engaged in hydrogen bonding. The experimental structure of solid ammonia is illustrated in Fig. 5. We notice that all three hydrogen atoms do, in fact, participate in hydrogen bonding, which means that the single lone pair has to accept no less than three hydrogen bonds.

The simple model of one lone pair per hydrogen bond is clearly not relevant here. The above discussion may be extended to include cases with more than one type of molecule in the structure. If the number of lone pairs available is less than the number of active hydrogen atoms we may encounter situations like that described for ammonia. In many other cases there is an excess of available lone pairs as for example in HF and in many organic compounds. Here, either some of the lone pairs are not involved in hydrogen bonding, or the donor hydrogen atom is directed towards some point between several lone pairs.

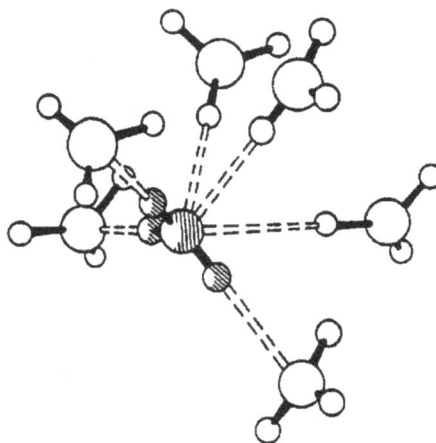

Fig. 5. The crystal structure of solid ammonia (Olovsson and Templeton [3], reprinted with permission, copyright 1959 IUCr).

From the above discussion it seems clear that the lone pairs should not be regarded as the immediate acceptors of hydrogen bonds, in the sense of "one lone-pair acceptor per H-bond donor". Many examples may be taken to illustrate that it is important to take the whole electron and nuclear distribution into account when discussing the relative arrangement of interacting molecules. The details of the hydrogen-bond arrangement are often mainly determined by simple geometrical requirements, like in ice, which will be discussed next.

4. The isolated water molecule and hexagonal ice Ih

One of the most interesting features in the crystal structure of hexagonal ice is the disorder of the orientation of the water molecules, as schematically illustrated in Fig. 6.

The lone pairs of the isolated water molecule are sometimes (even in standard textbooks) illustrated as "rabbit ears", as shown in Fig. 7. The tetrahedral distribution of four neighbours around the water molecules in ordinary ice is then often explained as due to the tetrahedral orientation of these lone pairs.

However, the actual electron distribution in the water molecule does not agree with this picture. As Fig. 8 shows, in the total and deformation density there is an even distribution in the non-bonded region, with no indication of concentration in the lone-pair directions. In the deformation density there is only a modest excess of electron density in the lone-pair directions very close to the oxygen nucleus. It is useful to remember that the sum of several sp^3-type hybrid orbitals results in a smooth electron density (spherically symmetric for

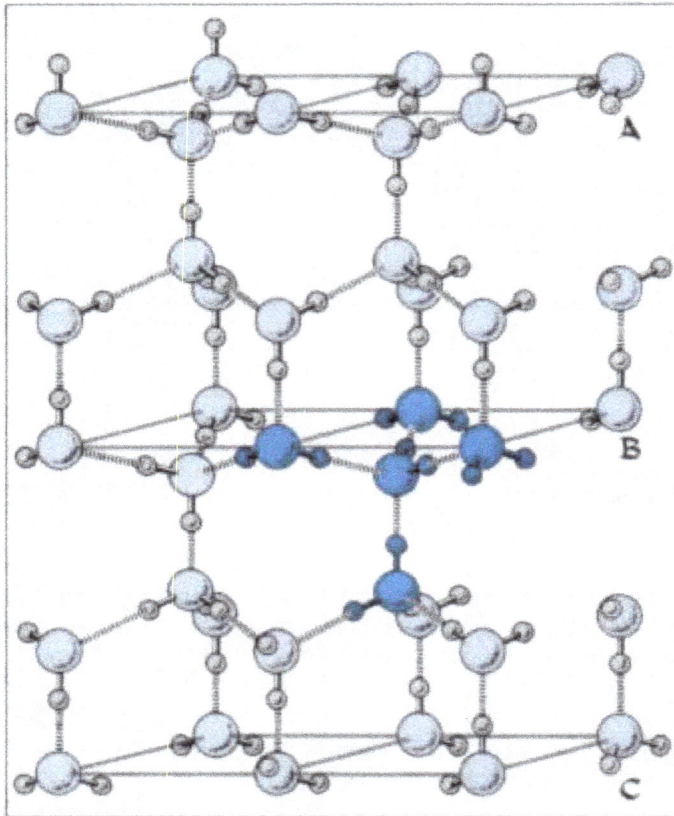

Fig. 6. The water molecules in ordinary hexagonal ice are disordered. The figure shows an example of possible orientations of the water molecules in adjacent unit cells (compare the molecules marked with A, B and C). If all possible orientations are statistically equally represented, an arrangement O-H---O and O---H-O is equally probable in one particular hydrogen bond. A diffraction study will then result in one half hydrogen atom at each place.

Fig. 7. Common illustration of the lone pairs in the water molecule.

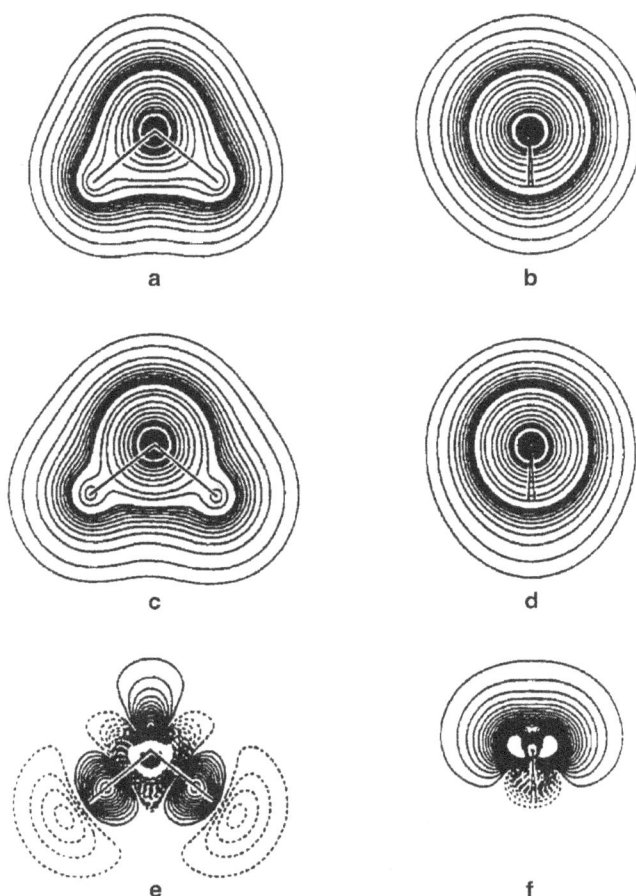

Fig. 8. The electron distribution in the free water molecule from *ab initio* quantum mechanical calculations (Hermansson [4]): a, b: The total electron density. c, d: The "promolecule" density (superposition of spherical atoms, before the molecule is formed). e, f: The deformation density (deformation as the molecule is formed, *i.e.* Fig. a minus Fig. c and Fig. b minus Fig. c). Figures a, c and e are shown in the plane of the water molecule, figures b, d and f perpendicular to this plane. Contour intervals in figures e and f are $\pm 0.05\,e/\text{Å}^3$ with cutoffs at $\pm 1.0\,e/\text{Å}^3$. Solid lines denote excess density, dashed lines deficiency.

four sp^3-orbitals). Moreover, molecular valence orbitals are of course different from atomic orbitals (and even more so when distributed by intermolecular interactions in a crystal, for example). We conclude that we should not expect to find traditional rabbit ear-like lone-pair densities for the free water molecule, nor for solid ice.

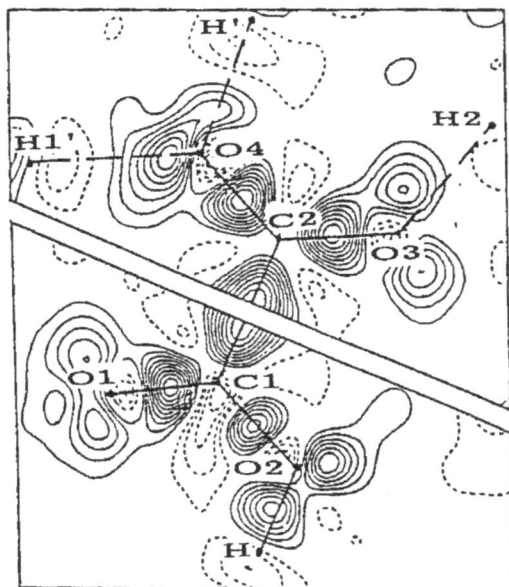

Fig. 9. $NaHC_2O_4 \cdot H_2O$. Experimental modelled deformation density of the $HC_2O_4^-$ ion (the ion is slightly twisted and the density is therefore shown in two different sections). Contours are drawn at intervals of $0.05\,e/Å^3$. Negative contours are dashed; the zero-level contour is omitted. (Delaplane, Tellgren and Olovsson [5], reprinted with permission, copyright 1990 IUCr).

The tetrahedral distribution in ice may be explained simply from the geometrical arrangement of the nuclei. As each water molecule donates two hydrogen bonds it also has to accept two hydrogen bonds (Fig. 4). The most favourable geometrical arrangement of the two acceptors is then realized if all four bonds are approximately tetrahedrally distributed. There is no need to refer to the model illustrated in Fig. 7, and as the actual electron distribution is quite different the "rabbit-ear model" should be avoided.

5. Other sp²- and sp³-hybridized oxygen atoms

Above it was concluded that the lone-pair maxima for the isolated water molecule are not very prominent, and that the H-bond arrangement in ice can be explained using geometrical arguments only. What is the situation for bound molecules involved in H---O bonds? Many examples *with* and *without* discernable electron-density excess in the traditional lone-pair directions of oxygen can be found in the literature. For example, many accurate experimental electron density studies of hydrates have been performed where electron density excess is indeed visible in the lone-pair region. One example is $NaHC_2O_4 \cdot H_2O$, shown in Fig. 9. Two distinct lobes in the deformation density are found around all the oxygen atoms. Again, the details of these lobes do not correspond to the electron-density distribution of "traditional" individual atomic lone-pairs. The lobes, or electron-density excess, observed should simply be considered as the net negative charge accumulation in the lone-pair region (sometimes called the non-bonded region). The directionality revealed in the scatter plots in Figs. 2 and 3 should accordingly be considered as a result of the general principle in hydrogen bond formation: *Hydrogen bonds are formed in directions of favourable negative charge density.* This principle should apply to all cases of weak and moderately strong hydrogen bonds where the electrostatic interaction is the dominating factor in determining the H-bond geometry. To summarize, it appears that for $NaHC_2O_4 \cdot H_2O$ the electron density excess in the "lone-pair directions" do have some directional influence on the incoming H-bonds, possibly combined with simple effects of the geometry of the acceptor molecule, in a similar way as was argued for water in ice above.

6. Effect of the environment on the electron density in the lone-pair region

An interesting question in this context is how much the negative charge distribution in the lone-pair region of the H-bond acceptor is modified by the interaction with the environment since this might affect the "H-bond directing ability" of the acceptor atom. This may be illustrated by the case of a water molecule involved in cation–water and/or hydrogen-bond interaction. Here pertinent experimental results are available from our studies of the electron density in hydrates of transition metal salts, such as $NiSO_4 \cdot 6H_2O$ and similar compounds [6–10].

The electron distribution in the lone-pair region of the three crystallographically independent water molecules is illustrated for $NiSO_4 \cdot 6H_2O$ [6] in Fig. 10. Water(1) is trigonally coordinated to nickel, water(2) is tetrahedrally coordinated to nickel in one of the lone-pair directions, water(3) is tetrahedrally coordinated to nickel in one of the lone-pair directions and accepts a hydrogen bond in the second direction.

The maps in Fig. 10 display the experimentally determined total electron density *minus* a superposition of free spherical neutral atoms, placed at the same positions as in the crystal. The maps thus show the electron redistribution due to both intra- and intermolecular bonding. The three water molecules in the $NiSO_4 \cdot 6H_2O$ crystal have quite similar geometries (r_{OH} all lie in the range 0.967–0.989 Å and the water angles in the range 109.5–111.2°). We can thus safely assume that a comparison of the maps in Fig. 10 will give us valuable information about the direct effect of the *intermolecular* interaction on the electron density distribution.

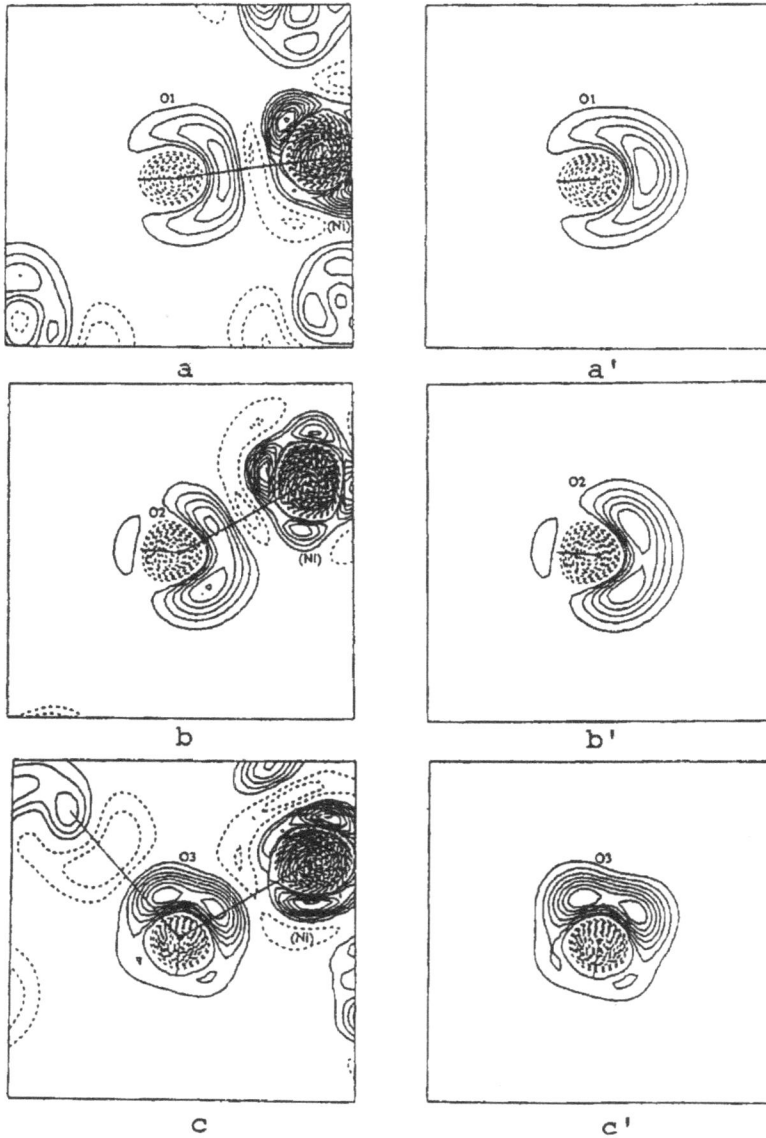

Fig. 10. Deformation electron density maps of the three water molecules in $NiSO_4 \cdot 6H_2O$, perpendicular to the water planes (in the "lone-pair" planes). The maps are model density maps from multipole refinements; in figures a′, b′ and c′ only the deformation functions of water have been plotted to eliminate "superposition effects" (*cf.* [7]). Contours are drawn at intervals of $0.05\ e\text{Å}^{-3}$. Negative contours are dashed; the zero-level contour is omitted. (Ptasiewicz-Bak, Olovsson and McIntyre [6], reprinted with permission, copyright 1993 IUCr).

Inspection of the maps in Fig. 10 shows that the uniform distribution of the deformation electron density in the lone-pair region of the free water molecule (Fig. 8f) is distorted very differently depending on the coordination situation. The distortion is probably predominantly a polarization effect and is just what might be expected from simple chemical arguments. Furthermore, we notice that the H-bonds seem to be directed towards the regions of highest electron density in the "lone-pair regions" of the three water molecules; this can represent quite different directions than the traditional lone-pair directions (*cf.* Fig. 10b). It is important to point out that some of the crystal-induced electron density excess in the direction of the incoming H-bond is caused by the H-bond itself, as can be seen for example from theoretical difference maps of the type '$\rho_{dimer} - \Sigma\rho_{molecules}$' for H-bonded dimers in the literature. Thus the rule that the H-bond is directed towards the region of large electron density is in part "a self-fulfilling statement".

The electron density maps in other similar compounds studied by us are in very good agreement with the above maps ($NiSO_4 \cdot 7H_2O$ [8], $Na_2Ni(CN)_4 \cdot 3H_2O$ [9], $NiCl_2 \cdot 4H_2O$ [10]. It thus appears to be possible to detect even the subtle effects on the electron density due to the environment. However, utmost precision in the experimental data is essential and collection at the lowest possible temperature.

It is relevant to make a few more comments on induced polarization and cooperative effects in the context of hydrogen bonding. When a water molecule donates a hydrogen bond to a neighbouring acceptor the polarity of the water molecule increases and the electron density is shifted towards the non-bonded region of oxygen. The donor water molecule will then become a better acceptor for another hydrogen bond, but at the same time it becomes a poorer donor of a second hydrogen bond. Correspondingly, the electron density of oxygen is shifted towards the non-bonded (acceptor) region if it receives a hydrogen bond, and the hydrogen becomes more positive and the water molecule becomes a better donor of a second hydrogen bond. Such cooperative effects occur in all coupled systems of hydrogen bonding and are of great importance, not least in biological systems where a very large number of hydrogen bonds may mutually influence each other. This is evidently the reason why the O---O hydrogen-bond distance is around 2.95 Å in the free water dimer but 2.76 Å in ordinary hexagonal ice with an extended network, where each water molecule is surrounded by four neighbours (Fig. 6). There are thus both short and long-range consequences of hydrogen bonding.

In liquid water there is a rapid exchange of water molecules and the number of neighbours varies with time; according to recent studies the distance between the neighbouring water molecules is around 2.75 Å, which, when comparing with ice Ih, suggests an average coordination close to four and a more ice-like structure than earlier believed. For recent high-quality X-ray scattering experiments on liquid water and comparison with earlier studies *cf.* Hura *et al.* [11], Sorensen *et al.* [12] and Hura *et al.* [13].

7. A critical analysis of geometry effects and lone-pair effects on the H-bond directionality for $(H_2O)_2$ and H_2O---OH^-

An important test case is the dimer of water, which has been studied in several extensive *ab initio* calculations as well as in experiments. The minimum energy occurs for a relative arrangement of the water molecules as illustrated in Fig. 11 (trans form, symmetry C_s), with $\varphi = 0°$ and an angle ε which according to various earlier calculations is approximately 50–55° (φ is the dihedral angle between the HOH plane of the donor molecule and the bisector plane of the acceptor molecule). This might suggest a clear directional influence of one of the lone pairs of the acceptor molecule. However, a more complete geometry analysis indicates that the deviation of ε from zero is also affected by other interactions, including repulsion between the non-bonded hydrogen atoms at the extremities of the dimer. Theoretical calculations of the dimer for different values of the ε and φ values have been presented earlier in the literature. The numbers for ΔE quoted below have been calculated by Müller [14]:

$$\varphi = 0°, \quad \varepsilon = -45°, \quad \Delta E = -3.92 \text{ kcal/mol}$$

$$\varphi = 0°, \quad \varepsilon = 0°, \quad \Delta E = -4.64 \text{ kcal/mol}$$

$$\varphi = 0°, \quad \varepsilon = +45°, \quad \Delta E = -5.03 \text{ kcal/mol}$$

$$\varphi = 90°, \quad \varepsilon = 0°, \quad \Delta E = -4.35 \text{ kcal/mol}$$

$$\varphi = 90°, \quad \varepsilon = +45°, \quad \Delta E = -4.25 \text{ kcal/mol}.$$

It is seen that when ε is varied between $+45°$ and $-45°$ for a fixed φ value of $0°$, the interaction energy changes by 1.1 kcal/mole. A large part of this can be attributed to difference in the electrostatic repulsion between the non-bonded hydrogen atoms on the two water molecules (using a nominal value of $+0.4e$ for the partial charge on hydrogen). Furthermore, when ε is kept fixed at $+45°$ and φ is rotated from $0°$ to $90°$, ΔE is seen to change by ~ 0.8 kcal/mole, although the incoming H-bond directionality is not changing. The energy change corresponds well to the change in the repulsive non-bonded H-H interaction.

The above examples may be taken as indications that there is no (or modest) directional influence of the individual lone pairs of the acceptor water molecule and that repulsion involving the hydrogen atoms not directly participating in the bonds plays a significant role in determining the geometry of the complex. In the cis-conformation ($\varepsilon = -45°$) the repulsion between the hydrogen atoms is maximum and the binding energy correspondingly smallest.

Extensive *ab initio* calculations have been performed for water-hydroxide ion clusters by Pliego and Riveros [17] and by Lee, Tarkeshwar and Kim [18]. The ΔE values below for the HOH---OH^- complex have been calculated

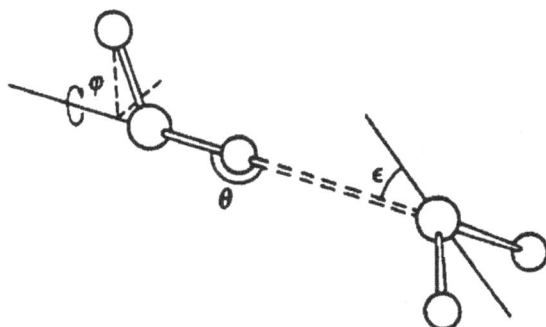

Fig. 11. The minimum-energy structure of the water dimer. The picture illustrates a φ value of $0°$, and an ε value of $45°$.

by Müller [14]); *cf.* Fig. 11 where the acceptor molecule has been replaced by OH⁻:

$$\varphi = 0°, \quad \varepsilon = 0°, \qquad \Delta E = -22.79 \, \text{kcal/mol}$$

$$\varphi = 0°, \quad \varepsilon = +45°, \qquad \Delta E = -24.72 \, \text{kcal/mol}$$

$$\varphi = 0°, \quad \varepsilon = +75°, \qquad \Delta E = -25.80 \, \text{kcal/mol}$$

$$\varphi = 0°, \quad \varepsilon = +90°, \qquad \Delta E = -25.26 \, \text{kcal/mol}.$$

The maximum bonding energy occurs for ε around $+75°$; the repulsion between the hydrogen atoms not directly participating in the hydrogen bond is then close to minimum.

However, the electron density in the hydroxide ion is different in comparison with water as shown in Fig. 12, and also the direction of maxima. It seems quite likely that for OH⁻ the lone-pair electron density excess sticking out at around $70°$ from the HO extension line plays a role in determining the geometry. This is supported by the fact that there are almost equally favourable energy minima for $\varepsilon + 70°$ and $\varepsilon = -70°$.

Our conclusion is that in the case of the water molecule there is an even distribution in the non-bonded region – no concentration in the lone-pair directions. Accordingly other factors, like repulsion, may dominate in determining the geometry of the dimer.

But in the case of the OH⁻ ion there is in contrast some concentration of the electron density in certain directions in the general lone-pair region and this may be the major directing factor in the HOH---OH⁻ complex. It may be remarked that it is the same number of electrons in HF as in OH⁻ and the angle between the electron density lobes and the HF extension line is similar, as shown in Fig. 12. This may explain the observed as well as calculated bent geometry of the HF dimer and other HF complexes, as is well known from the literature [19, 20]. The geometry of the HOH---OH⁻ complex again demon-

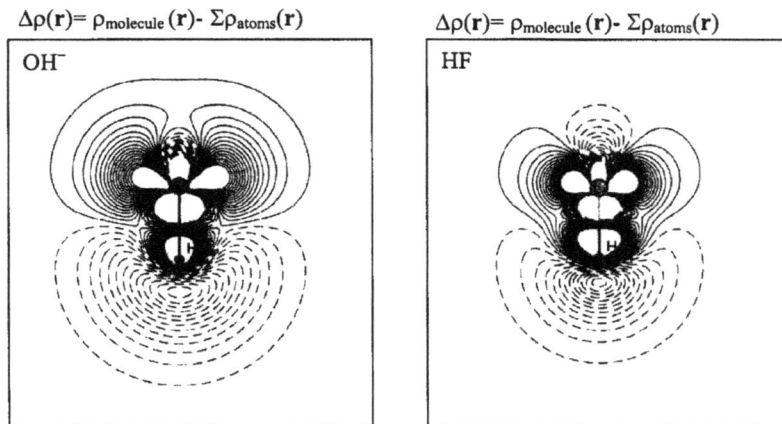

Fig. 12. Static theoretical deformation density in the free OH^- ion and the HF molecule. Optimized structures from 6-31G**/HF calculations by Müller [14]. r(OH) = 0.9850 Å, r(HF) = 0.9000 Å. Contour levels at $\pm 0.020\,e/\text{Å}^3$ with cutoffs at $\pm 0.34\,e/\text{Å}^3$. Zero contour omitted.

strates the general principle: Hydrogen bonds are directed towards regions of favourable negative charge density.

In the discussion above it has been suggested that repulsion between the hydrogen atoms, plays an important role for the geometry when there is no concentration of density in certain directions. But no proof is presented that this is the dominating effect in the water dimer. On the other hand in the case of the OH^- ion the lobes in the theoretical electron density may play a role in determining the geometry. Very accurate determinations of the details of the charge distribution in the molecules, as well as appropriate energy decomposition schemes, are needed to evaluate the relative importance of other factors, such as electrostatic forces between the different multipoles for the geometry of approach.

When arguing about the possible role of repulsion between hydrogen atoms in the water dimer it is interesting to study the results summarized by Burnham and Xantheas [21]. In their Fig. 11 the acceptor angle ε is plotted as function of the O-O separation for various models. There is a drastic increase in the acceptor angle when the O-O separation decreases: For the ASP-W model from 60° to 110° at 2.85 Å, for the ASP-W4 model from 50° to 110° at 2.70 Å! Although these results are very strongly model dependent it is tempting to suggest that the repulsion between the hydrogen atoms plays a role here.

Finally it should be remarked that the water molecule may function as a donor as well as acceptor and it is accordingly rather unique from a chemical bonding point of view. There are thus several reasons for the important function of water in biological as well as in inorganic and organic systems.

The Role of the Lone Pairs in Hydrogen Bonding   **977**

8. Conclusions

From structural data it can be concluded that the immediate acceptor of a hydrogen bond is some negative charge accumulation, such as in a lone-pair region, *but not specifically any individual lone pairs in the traditional, atomic sense.* This is most likely true for all weak and moderately strong hydrogen bonds for which the major factor determining the hydrogen-bond geometry is electrostatic interaction. In the case of very strong hydrogen bonds covalent contributions may explain deviations from this picture. The water dimer is often taken as a typical example of the directional influence of the lone pairs. However, theoretical studies of the water dimer with different angles of approach suggest that the trans configuration of the water dimer is mainly a result of other interactions, namely repulsion between those hydrogen atoms of the donor and acceptor molecules which are not participating in the hydrogen bond. For the H_2O---OH^- complex also the lone-pair region affects the directionality.

Acknowledgement

I wish to thank Carsten Müller for theoretical calculations and Prof. Kersti Hermansson and Dr. Daniel Spångberg for valuable discussions. This work has been supported by grants from the Trygger Foundation for Scientific Research.

References

1. J. Kroon, J. A. Kanters, J. G. C. M. van Duijneveldt-van de Rijdt, F. B. van Duijneveldt, and J. A. Vliegenhart, J. Mol. Struct. **24** (1975) 109.
2. P. Murray-Rust and J. P. Glusker, J. Am. Chem. Soc. **106** (1984) 1018.
3. I. Olovsson and D. H. Templeton, Acta Cryst. **12** (1959) 832.
4. K. Hermansson, Acta Univ. Upsaliensis No. 744 (thesis) (1984).
5. R. G. Delaplane, R. Tellgren, and I. Olovsson, Acta Cryst. **B46** (1990) 361.
6. H. Ptasiewicz-Bak, I. Olovsson, and G. J. McIntyre, Acta Cryst. **B49** (1993) 192.
7. I. Olovsson, H. Ptasiewicz-Bak, and G. J. McIntyre, Z. Naturforschung **48a** (1993) 3.
8. H. Ptasiewicz-Bak, I. Olovsson, and G. J. McIntyre, Acta Cryst. **B53** (1997) 325.
9. H. Ptasiewicz-Bak, I. Olovsson, and G. J. McIntyre, Acta Cryst. **B54** (1998) 600.
10. H. Ptasiewicz-Bak, I. Olovsson, and G. J. McIntyre, Acta Cryst. **B55** (1999) 830.
11. G. Hura, J. M. Sorensen, R. M. Glaeser, and T. Head-Gordon, J. Chem. Phys. **113** (2000) 9140.
12. J. M. Sorensen, G. Hura, R. M. Glaeser, and T. Head-Gordon, J. Chem. Phys. **113** (2000) 9149.
13. G. Hura, D. Russo, R. M. Glaeser, T. Head-Gordon, M. Krack, and M. Parrinello, Phys. Chem. Chem. Phys. **5** (2003) 1981.
14. C. Müller, Private communication. The data were calculated using the program Gaussian 03 [15], Hartree–Fock/6-31G**, correction for basis set superposition errors; geometry fixed to O-H = 0.97 Å, O---O = 2.92 Å, H-O-H = 104.5°. Binding energies calculated with counter-poise method, Boys and Bernadi [16]).

15. Gaussian 03, Revision C.02, M. J. Frisch, G. W. Trucks, H. B. Schlegel,
    G. E. Scuseria, M. A. Robb, J. R. Cheeseman, J. A. Montgomery, Jr., T. Vreven,
    K. N. Kudin, J. C. Burant, J. M. Millam, S. S. Iyengar, J. Tomasi, V. Barone, B. Men-
    nucci, M. Cossi, G. Scalmani, N. Rega, G. A. Petersson, H. Nakatsuji, M. Hada,
    M. Ehara, K. Toyota, R. Fukuda, J. Hasegawa, M. Ishida, T. Nakajima, Y. Honda,
    O. Kitao, H. Nakai, M. Klene, X. Li, J. E. Knox, H. P. Hratchian, J. B. Cross,
    C. Adamo, J. Jaramillo, R. Gomperts, R. E. Stratmann, O. Yazyev, A. J. Austin,
    R. Cammi, C. Pomelli, J. W. Ochterski, P. Y. Ayala, K. Morokuma, G. A. Voth,
    P. Salvador, J. J. Dannenberg, V. G. Zakrzewski, S. Dapprich, A. D. Daniels,
    M. C. Strain, O. Farkas, D. K. Malick, A. D. Rabuck, K. Raghavachari, J. B. Fores-
    man, J. V. Ortiz, Q. Cui, A. G. Baboul, S. Clifford, J. Cioslowski, B. B. Stefanov,
    G. Liu, A. Liashenko, P. Piskorz, I. Komaromi, R. L. Martin, D. J. Fox, T. Keith,
    M. A. Al-Laham, C. Y. Peng, A. Nanayakkara, M. Challacombe, P. M. W. Gill,
    B. Johnson, W. Chen, M. W. Wong, C. Gonzalez, and J. A. Pople. Gaussian, Inc.,
    Wallingford CT (2004).
16. S. F. Boys and F. Bernadi, Mol. Phys. **19** (1970) 553.
17. J. R. Pliego and J. M. Riveros, J. Chem. Phys. **112** (2000) 4045.
18. H. M. Lee, P. Tarkeshwar, and K. S. Kim, J. Chem. Phys. **121** (2004) 4657.
19. T. R. Dyke, B. J. Howard, and W. Klemperer, J. Chem. Phys. **56** (1972) 2442.
20. K. C. Janda, J. M. Steed, S. E. Nowick, and W. Klemperer, J. Chem. Phys. **67** (1977)
    5162.
21. C. J. Burnham and S. S. Xantheas, J. Chem. Phys. **116** (2002) 1479.

Z. Phys. Chem. **220** (2006) 797–810 / **DOI** 10.1524/zpch.2006.220.7.797
© by Oldenbourg Wissenschaftsverlag, München

Comparison of the Proton Transfer Path in Hydrogen Bonds from Theoretical Potential Energy Surfaces and the Concept of Conservation of Bond Order

By Ivar Olovsson*

Department of Materials Chemistry, Ångström Laboratory, Box 538,
SE-751 21 Uppsala, Sweden

*Dedicated to Prof. Dr. Dr. h.c. Wolf Weyrich on the occasion
of his 65ʰ birthday*

(Received February 22, 2006; accepted March 2, 2006)

*Hydrogen Bonds / Proton Transfer Path / Potential Energy Surface /
Reaction Coordinates / Pauling Bond Order*

The 'quantum-mechanically derived reaction coordinates' (QMRC) for the proton transfer in hydrogen bonds involving fluorine, oxygen and chlorine have been derived from earlier *ab initio* calculations of potential energy surfaces. A comparison is made between QMRC and the corresponding reaction coordinates (BORC) derived by applying the Pauling bond order concept together with the principle of conservation of bond order. Theoretical calculations have shown that the sum of the bond orders remains close to constant along the reaction coordinate in agreement with the Pauling postulate. The BORC correlation curves agree very well with theoretical results. The results indicate that the BORC curve gives a good representation of the reaction coordinates (proton transfer path) for any X-H---Y aggregate.

1. Introduction

As an increasing number of crystallographic data becomes available the interest in studying systematic trends in various structural features in crystals has strongly increased. Hydrogen bonds play an important role in many chemical systems and are vital in biological processes and studies of their general properties are accordingly important. Hydrogen bonds are particularly effective in a search for systematic trends for several reasons: (1) In hydrogen bonds

---

* Corresponding author. E-mail: ivar.olovsson@mkem.uu.se

RX-H---YR' the groups R and R' bonded to the donor and acceptor atoms X and Y may be varied within very wide limits. This makes it possible to systematically study series of different types of compounds with identical bonds X-H---Y. (2) The characteristic distances X-H, H---Y and X---Y vary over a very large range. For example, in a O-H---O bond the O-H distance varies from 0.96 Å, the covalent distance in the free donor molecule, to ~ 2.40 Å, when the proton has been transferred to the acceptor molecule and is in van der Waals contact with the original donor oxygen atom.

In a series of different crystalline compounds, where the same atoms X and Y are involved in the X-H---Y hydrogen bond, the distribution of the proton positions observed will be a result of pertubations of different magnitudes by the crystalline environment. However, the assembly of such points may be assumed to be distributed close to the minimum energy path for the transfer of a proton along this particular kind of hydrogen bond of the isolated system. This is the assumption behind the mapping of chemical reaction pathways from crystal structure data, and which has been applied to a variety of crystal structures by Dunitz and collaborators, *cf.* [1, 2]. Based on this assumption, the ideal correlation curve for a particular type of X-H---Y hydrogen bond can evidently be chosen as the reaction coordinates derived from the theoretical potential energy surface of the system in question.

As well known, by applying the Pauling bond order concept together with the principle of conservation of bond order a simple functional representation may be derived for the relation between the X-H and H---Y distances (as well as for the interdependence of the distances in other linear triatomic systems, *cf.* [1].) However, there is evidently no theoretical basis for a functional relation such as that proposed by Pauling. Considering the popular use of the bond order concept it is of considerable interest compare the quantum-mechanically derived reaction coordinates (QMRC) with the corresponding reaction coordinates derived by the Pauling approach (called BORC in the following).

The purpose of the present paper is to test whether the principles used in applying the Pauling bond order concept are in reasonable agreement with the results from theoretical calculations:

(1) Does the sum of the bond orders for X--H and H--Y, respectively, remain constant along the reaction coordinate also according to theoretical calculations?

(2) How do the bond orders derived from theoretical calculations compare with the Pauling bond order?

2. Reaction coordinates from theoretical potential energy surface, QMRC

The reaction coordinates for the proton transfer in three hydrogen-bonded systems involving fluorine, oxygen and chlorine have been derived from

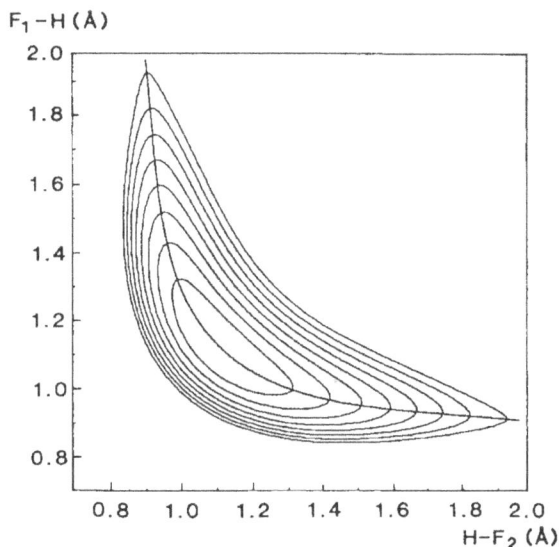

Fig. 1. Potential energy surface and reaction coordinates for the proton transfer in the $(F_1\text{--}H\text{--}F_2)^-$ ion. Reproduced by Olovsson and Jaskolski [6] from *ab initio* SCF calculations by Almlöf [3].

some earlier *ab initio* calculations of potential energy surfaces. They are based on the following MO LCAO SCF calculations: on HF_2^- by Almlöf [3], on $H_3O_2^-$ by Stögård, Strich, Almlöf and Roos [4] and on HCl_2^- by Almlöf [5]. In all cases the hydrogen bonds are assumed to be linear. In these papers the potential energy surface has been expressed in terms of Q_1 and Q_2 where Q_1 denotes the distance $X(1)\text{---}X(2)$ and Q_2 the difference between the $X(1)\text{-}H$ and $H\text{-}X(2)$ distances. For present purpose the potential energy surface has instead been expressed in terms of $X(1)\text{-}H$ and $H\text{-}X(2)$. The results for HF_2^- and HCl_2^- are shown in Figs. 1 and 2. The minimum energy proton transfer path, corresponding to the gradient of the potential energy surface, is also indicated in these figures. In the following it will be denoted by QMRC (**Q**uantum **M**echanical **R**eaction **C**oordinate). The details of the procedure to derive the minimum energy proton transfer path have been described in an earlier paper by Olovsson and Jaskolski [6].

In the case of $H_3O_2^-$ it was not possible to replot the potential energy surface as the coefficients in the polynomial expression were not published. In view of the very great importance of O-H---O bonds (by far the largest number of hydrogen bonds investigated) we have traced the reaction coordinate from the contour level diagrams of the paper by Stögård *et al.* [4]. The curve thus obtained is shown in Fig. 3. The general shape of the curves in Figs. 1, 2 and 3 is

Cl_1-H (Å)

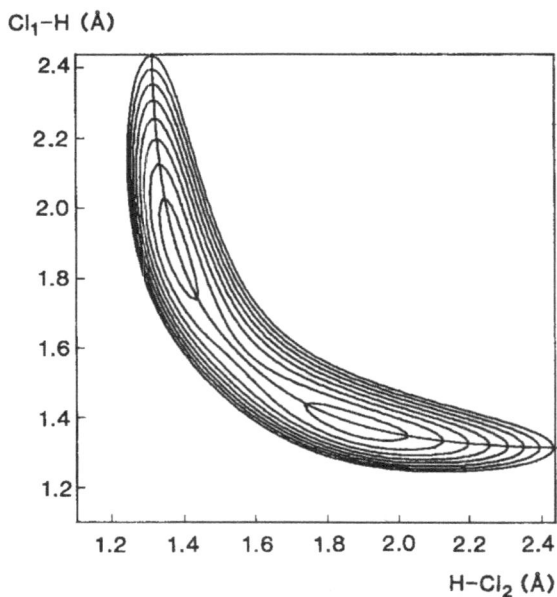

H–Cl_2 (Å)

Fig. 2. Potential energy surface and reaction coordiates for the proton transfer in the $(Cl_1\text{--}H\text{--}Cl_2)^-$ ion. Reproduced by Olovsson and Jaskolski [6] from *ab initio* SCF calculations by Almlöf [5].

O_1-H (Å)

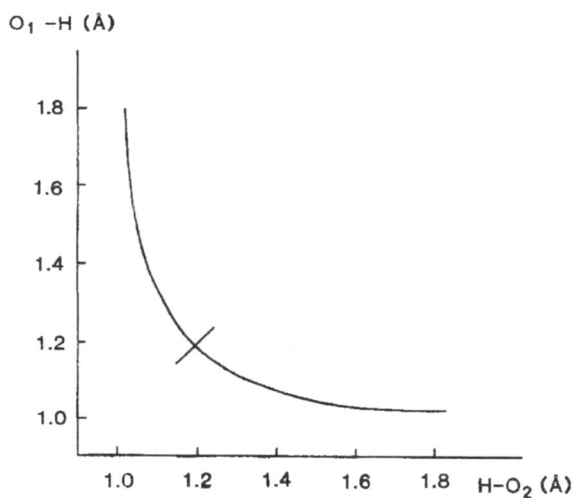

H–O_2 (Å)

Fig. 3. Reaction coordinates for the proton transfer in the $(HO_1\text{--}H\text{--}O_2H)^-$ ion. Traced by Olovsson and Jaskolski [6] from the potential energy surface obtained in *ab initio* SCF calculations by Stögård *et al.* [4].

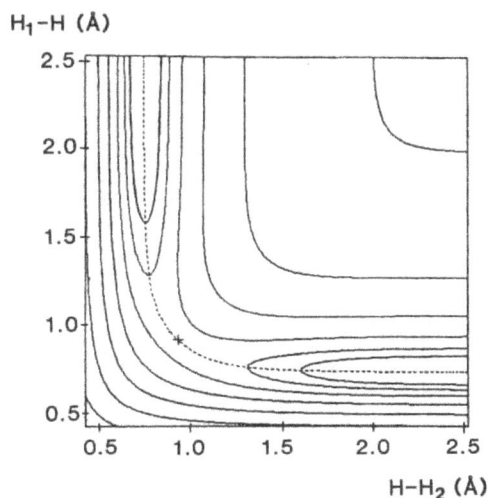

Fig. 4. Potential energy surface and reaction coordiates for the proton transfer in (H-H---H). From Liu [7], Siegbahn and Liu [8], Truhlar and Horowitz [9].

completely analogous. On an absolute scale they must differ as these hydrogen bonds have different characteristic lengths.

Although H_3 does not represent a conventional H-bonded system, it is formally completely analogous, and it has been included here in order to check the effectiveness of the true minimum energy reaction path. The H_3 system has been extensively studied as an example of the simplest triatomic transfer reaction: $H-H + H \rightarrow H--H--H \rightarrow H + H-H$. The complete potential energy surface has been very accurately calculated by Liu [7] and Siegbahn and Liu [8] using the configuration interaction (CI) method; a detailed functional representation of this surface has subseqently been published by Truhlar and Horowitz [9]. From these theoretical results we can thus make a comparison with what is believed to be the true minimum energy reaction path, Fig. 4.

3. Reaction coordinates from bond orders, BORC

The concept of bond order has been popular in chemistry for a very long time and is often quite useful in 'explaining' systematic trends in the bond lengths in related compounds. Already in 1947 Pauling [10] introduced the idea that there is a simple relation between bond length and 'bond order' or 'valence'. Several expressions for this relation have been proposed [11] but the most popular is still due to Pauling:

$$d(\rho) - d(1) = \Delta d = -a \ln \rho$$

where $d(\rho)$ is the interatomic distance for a fractional bond with bond order ρ and $d(1)$ is the corresponding single bond length. In a transfer reaction X-H + Y $\rightarrow$ X--H--Y $\rightarrow$ X + H-Y, it is postulated that the sum (n) of the bond orders ρ_1 and ρ_2 for X--H and H--Y, respectively, will remain constant along the reaction coordinate. This is also the basic assumption in the so-called Bond Energy Bond Order (BEBO) method in the theory of chemical reactions; *cf.* Johnston [12], Johnston and Parr [13], Agmon [14, 15]. We thus obtain:

$$\exp(-\Delta d_1/a_1) + \exp(-\Delta d_2/a_2) = n \qquad (1)$$

For the reactants, $d(\rho_1) = d$(X-H, free) and $d(\rho_2) = d$(H---Y, ∞), so that $\rho_1 = 1$ and $\rho_2 = 0$; for the products $\rho_1 = 0$, $\rho_2 = 1$. In a hydrogen bond the bond order of the two chemical bonds X-H and H---Y must add up to 1, so that $\rho_1 + \rho_2 = 1$ all along the reaction coordinate. The form of this curve is thus obtained from the Pauling relation under the condition that $\rho_1 + \rho_2 = 1$. We will refer to the curve thus calculated as the Bond Order Reaction Coordinate (BORC). In the following we will mainly discuss systems for which X = Y, in which case $a_1 = a_2$.

Is it reasonable to assume that the sum of the bond orders remains constant along the reaction coordinate? To check this *ab initio* quantum mechanical calculations have been made for the [F-H---F]$^-$ and [HO-H---OH]$^-$ systems using Gaussian 03 [16, 17]. From these calculations different 'theoretical bond orders' may be derived, although the definition of the bond order varies. For the [F-H---F]$^-$ system the Wiberg bond order [18] varies from 0.645 to 0.660 (as F-H varies from 0.92 to 1.12 Å). For the [HO-H---OH]$^-$ system the bond order is in the range 0.695–0.720. These results may be taken as justification for the above assumption that the sum of the bond order remains constant along the reaction coordinate, although the sum differs from one. Note also that this sum is not the same for the two systems tested (whereas it is assumed to be the same, and equal to one, in all cases in the Pauling approach). It may be remarked that the theoretically derived bond orders depend not only on their definition but also on the partitioning of the electron density.

The values of a in the Pauling relation have been derived from the following bond lengths (Steiner and Saenger [19] have used sligthly different bond lengths as reference resulting in somewhat different a-values; *cf.* also Steiner [20]).

3.1 System F-H---F

$d(1) = 0.917$ Å, the spectroscopically determined equilibrium value (r_e) for the F-H distance in gaseous hydrogen fluoride (Herzberg [21]), $d(0.5) = 1.123$ Å, half the distance in a symmetrical F--H--F bond, for which $\rho_1 = \rho_2 = 0.5$. The F---F distance has then been set equal to 2.246 Å, which is the minimum distance in the theoretical calculationon HF$_2{}^-$ by Almlöf [3]. These two bond distances give $a(\mathrm{F}) = 0.297$ Å. (The experimental values found for the $\mathrm{HF_2}^-$ ion in crystal structures of alkali metal salts are around 2.26 Å, Ibers [22].)

3.2 System O-H---O

$d(1) = 0.957$ Å, the spectroscopically determined equilibrium value (r_e) for the O-H distance in the free water molecule (Benedict, Gailar and Plyler [23]), $d(0.5) = 1.200$ Å, half the distance in a symmetrical O-H---O bond, which has been set equal to 2.400 Å; this is considered to be a representative experimental value for the shortest bonds of this type (*cf.* Olovsson and Jönsson [24], Joswig, Fuess and Ferraris [25]). These values give $a(\mathrm{O}) = 0.351$ Å.

3.3 System N-H---N

$d(1) = 1.014$ Å, the spectroscopically determined equilibrium N-H distance in gaseous ammonia (*cf.* Herzberg [21]). The value of $d(0.5)$ is much more uncertain in this case. The hydrogen bonds N-H---N are in general relatively long and strongly asymmetric. From general, qualitative arguments it may be concluded that the shortest hydrogen bonds will probably have a N---N distance around 2.60 Å (Olovsson and Jönsson [24]). Thus, we will here set $d(0.5) = 1.30$ Å. The above values give $a(\mathrm{N}) = 0.413$ Å.

3.4 System Cl-H---Cl

$d(1) = 1.275$ Å, the spectroscopically determined equilibrium H-Cl distance in gaseous hydrogen chloride (Herzberg [26]). The value of $d(0.5)$ is also in this case rather uncertain. In the theoretical calculations for the $\mathrm{HCl_2}^-$ ion by Almlöf [5], a minimum Cl---Cl distance of 3.130 Å was obtained. We will thus set $d(0.5) = 1.565$ Å. These values give $a = 0.418$ Å.

3.5 Systems X-H---Y

The curves for systems with X different from Y may be obtained quite simply by inserting $a(1) = a(\mathrm{X})$ and $a(2) = a(\mathrm{Y})$ in Eq. (1).

3.6 Trimeric hydrogen, H-H---H

In the calculation of the BORC curve we have chosen $d(1) = 0.741$ Å (1.4008 a.u.), the spectroscopically determined equilibrium value (Stoicheff [27, 28]). For this system no experimental value for the distance in a "symmetrical hydrogen bond" is known. Here we will adopt $d(0.5) = 0.930$ Å (1.757 a.u.) the theoretically obtained H..H distance at the saddle point for symmetrical H--H--H (Liu [7]). These values give $a(\mathrm{H}) = 0.272$ Å.

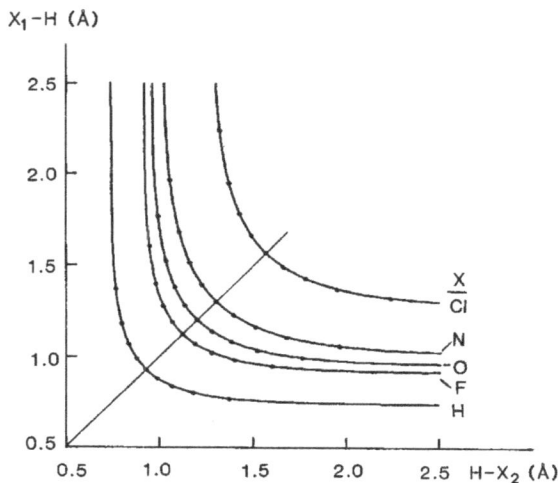

Fig. 5. Bond order reaction coordinates (BORC) calculated for (H-H---H), (F-H..F), (O-H---O), (N-H---N) and (Cl-H---Cl) bonds (*cf.* text).

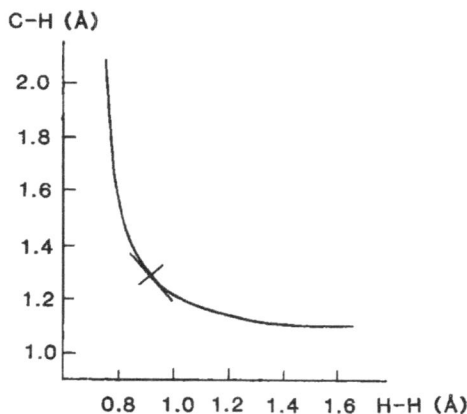

Fig. 6. Reaction coordinates calculated by the 'bond energy bond order' (BEBO) method by Johnston [13] for the reaction $CH_3 + H_2 \rightarrow CH_4 + H$.

With the a-values calculated above, the BORC curves for H-H---H, F-H---F, O-H---O, N-H---N and Cl-H---Cl bonds are shown in Fig. 5. The same principle was also applied by Johnston [10] in the "bond energy bond order" (BEBO) method for the reaction $CH_3 + H_2 \rightarrow CH_4 + H$. We notice that this reaction path (Fig. 6) is completely analogous to the reaction paths shown above for hydrogen-bonded systems.

Table 1. Reaction coordinates for H(1)--H(2)--H(3).

H(1)--H(2) (Å)	H(2)--H(3) (Å)	
	QMRC	BORC
[0.930]	[0.930]	[0.930]
0.965	0.898	0.898
1.008	0.867	0.869
1.054	0.841	0.845
1.103	0.820	0.825
1.153	0.803	0.809
1.204	0.791	0.796
1.256	0.781	0.786
1.309	0.773	0.777
1.361	0.766	0.771
1.414	0.761	0.765
1.466	0.757	0.761
1.519	0.754	0.757
1.572	0.752	0.754
1.625	0.750	0.752
1.678	0.748	0.750
1.731	0.747	0.749
1.784	0.746	0.747
1.837	0.745	0.746
1.890	0.745	0.745
[∞]	–	[0.741]

4. Comparison between QMRC and BORC

4.1 Trimeric hydrogen, H-H---H

The minimum energy reaction path QMRC obtained in the theoretical calculations of the potential energy surface for H_3 (Fig. 4) is apparently quite well represented by the BORC curve (Fig. 5), see Table 1. The comparison naturally only concerns the general form of the BORC curve as the two extreme points H(1)-H(2) = H(2)-H(3) = 0.741 Å and H(1)-H(2) = H(2)-H(3) = 0.930 Å are just the same in the two cases. The difference between the QMRC and BORC values is less than 0.006 Å, which is estimated to be of the same order of magnitude as the combined uncertainty due to errors in the parameter a, the quantum mechanical results and the plotting of the reaction coordinate.

4.2 F-H---F system

In this case, a more conventional H-bonded system, there is an even closer fit between the QMRC and BORC curves (Figs. 1 and 5), see Table 2. Apart from the first two values (where the polymomial expansion applied in the derivation of QMRC is less precise), the values agree within 0.001 Å. However, the absolute uncertainty of the quantum mechanical results is at least an order of magnitude larger.

Table 2. Reaction coordinates for F(1)--H--F(2).

| F(1)--H (Å) | H--F(2) | |
	QMRC	BORC
[0.917]	–	[∞]
0.930	(1.794)	1.854
0.940	(1.685)	1.689
0.950	1.588	1.587
0.960	1.514	1.513
0.970	1.456	1.456
0.980	1.410	1.409
0.990	1.370	1.370
1.000	1.337	1.337
1.010	1.308	1.307
1.020	1.283	1.282
1.030	1.260	1.259
1.040	1.239	1.239
1.050	1.220	1.220
1.060	1.203	1.203
1.070	1.188	1.188
1.080	1.174	1.173
1.190	1.160	1.160
1.100	1.148	1.148
1.110	1.137	1.137
1.120	1.126	1.126
[1.123]	[1.123]	[1.123]
[∞]	–	[0.917]

Table 3. Reaction coordinates for O(1)--H--O(2).

| O(1)--H (Å) | H--O(2) (Å) | |
	QMRC	BORC
[0.957]	–	[∞]
1.02	(1.75)	1.59
1.03	(1.60)	1.54
1.04	1.52	1.50
1.05	1.48	1.47
1.06	1.44	1.44
1.07	1.41	1.41
1.08	1.38	1.38
1.09	1.35	1.36
1.10	1.33	1.34
1.11	1.30	1.32
1.12	1.28	1.30
1.13	1.27	1.29
1.14	1.25	1.27
1.15	1.24	1.26
1.16	1.22	1.24
1.17	1.21	1.23
1.18	1.20	1.22
1.19	1.19	1.21
1.20	1.18	1.20
[∞]	–	[0.957]

Table 4. Reaction coordinates for Cl(1)--H--Cl(2).

Cl(1)--H (Å)	H--Cl(2) (Å)	
	QMRC	BORC
[1.275]	–	[∞]
1.330	2.132	2.151
1.340	2.058	2.086
1.350	2.000	2.031
1.360	1.953	1.984
1.370	1.913	1.942
1.380	1.878	1.905
1.390	1.848	1.872
1.400	1.821	1.841
1.410	1.796	1.814
1.420	1.774	1.789
1.430	1.753	1.766
1.440	1.734	1.744
1.450	1.716	1.724
1.460	1.699	1.706
1.470	1.684	1.688
1.480	1.669	1.672
1.490	1.655	1.656
1.500	1.641	1.642
1.510	1.628	1.628
1.520	1.616	1.615
1.530	1.604	1.603
1.540	1.593	1.592
1.550	1.582	1.581
1.560	1.572	1.570
1.570	1.562	1.560
[∞]	–	[1.275]

4.3 O-H---O system

The fit between the QMRC and BORC curves (Figs. 3 and 5) is illustrated in Table 3. As remarked earlier, QMRC was traced directly from the previously published contour level diagram. From a comparison with the HF_2^- results it seems likely that the disagreement between the QMRC and BORC values in Table 3 (≤ 0.02 Å) is partly due to such tracing errors. At the same time the absolute uncertainty in these early quantum mechanical calculations is also of the same order of magnitude.

4.4 Cl-H---Cl system

The quantitative comparison between the QMRC and BORC curves (Figs. 2 and 5) is shown in Table 4. It seems likely that the disagreement (≤ 0.03 Å) is to a large extent due to the deficiency in the polynomial expansion applied in

Table 5. Reaction coordinates, BORC, for N(1)--H--N(2).

N(1)--H (Å)	H--N(2)
1.014	[∞]
1.050	2.038
1.060	1.942
1.070	1.866
1.080	1.803
1.090	1.749
1.100	1.703
1.110	1.663
1.120	1.627
1.130	1.594
1.140	1.565
1.150	1.538
1.160	1.513
1.170	1.491
1.180	1.470
1.190	1.450
1.200	1.432
1.210	1.415
1.220	1.399
1.230	1.384
1.240	1.370
1.250	1.357
1.260	1.344
1.270	1.332
1.280	1.321
1.290	1.310
1.300	1.300
[∞]	[1.014]

the derivation of the QMRC curve for values far from the center of the curve (*cf.* the F-H---F system).

4.5 Other X-H---Y bonds

An analogous comparison between the reaction path derived from a theoretically calculated potential energy surface and the BORC curve cannot be done for other X-H---Y bonds. Particularly important should be a study of the N-H---N system. For completeness, the numerical values corresponding to the BORC curve for this system (Fig. 5) are given in Table 5.

In principle there is no reason to believe that the agreement between QMRC and BORC would be less good in other cases when sufficiently accurate and complete quantum mechanical results are available. However, it should be emphasized that the above comparison between QMRC and BORC has only tested the effectiveness of the functional relation given in Eq. (1) to represent the interdependence of the X-H and H-Y distances.

5. Conclusions

Theoretical calculations have demonstrated that the sum of the bond orders remains close to constant along the reaction coordinate in agreement with the Pauling postulate. The correlation curves (BORC) derived by assuming that the sum of the bond orders is equal to one are in close agreement with theoretical results. The results indicate that the BORC curve gives a good representation of the reaction coordinates (proton transfer path) for any X-H---Y aggregate.

Acknowledgement

Thanks to Carsten Müller for theoretical calculations of bond orders and to Prof. Kersti Hermansson for useful discussions. This work had been supported by a grant from the Trygger Foundation for Scientific Research.

My close contacts with Wolf Weyrich date back to 1982-83 when I was invited as guest professor to the university of Konstanz and spent one year in his laboratory (as well as in the bottom floor of his house – as far as I know without any complaints from the family members). At that time I participated in a project involving studies of hydrogen bonding in single crystals of $KHCO_3$ by Compton scattering. 31st May 1991 I had the pleasure as faculty promotor of the university of Uppsala to confer the degree of *doctor honoris causa* to Wolf Weyrich. In 1999 I again spent three months in Konstanz, at this time taking part in theoretical studies of hexagonal ice – due to its disordered structure a formidable problem. As hydrogen bonding is a field of great interest to both of us it is a pleasure for me to dedicate this paper to Wolf Weyrich on his 65th birthday.

References

1. J. D. Dunitz, *X-Ray Analysis and the Structure of Organic Molecules*. Cornell Univ. Press, Ithaca, NY (1979).
2. H. B. Bürgi and J. D. Dunitz, Acc. Chem. Res. **16** (1983) 153.
3. J. Almlöf, Chem. Phys. Letters **17** (1972) 49.
4. A. Stögård, A. Strich, J. Almlöf, and B. Roos, Chem. Phys. **8** (1975) 405.
5. J. Almlöf, J. Mol. Struct. **85** (1981) 179.
6. I. Olovsson and M. Jaskolski, Polish J. Chem. **60** (1986) 759.
7. B. Liu, J. Chem. Phys. **58** (1973) 1925.
8. P. Siegbahn and B. Liu, J. Chem. Phys. **68** (1978) 2457.
9. D. G. Truhlar and C. J. Horowitz, J. Chem. Phys. **68** (1978) 2466.
10. L. Pauling, J. Am. Chem. Soc. **69** (1947) 542.
11. I. D. Brown, Acta Cryst. B **48** (1992) 553.
12. H. S. Johnston, Adv. Chem. Phys. **3** (1960) 131.
13. H. S. Johnston and C. A. Parr, J. Am. Chem. Soc. **85** (1963) 2544.
14. N. Agmon, Chem. Phys. Letters **45** (1977) 343.
15. N. Agmon, J. Chem. Soc. Faraday II **74** (1978) 388.

16. Gaussian 03, Revision C.02, M. J. Frisch, G. W. Trucks, H. B. Schlegel, G. E. Scuseria, M. A. Robb, J. R. Cheeseman, J. A. Montgomery, Jr., T. Vreven, K. N. Kudin, J. C. Burant, J. M. Millam, S. S. Iyengar, J. Tomasi, V. Barone, B. Mennucci, M. Cossi, G. Scalmani, N. Rega, G. A. Petersson, H. Nakatsuji, M. Hada, M. Ehara, K. Toyota, R. Fukuda, J. Hasegawa, M. Ishida, T. Nakajima, Y. Honda, O. Kitao, H. Nakai, M. Klene, X. Li, J. E. Knox, H. P. Hratchian, J. B. Cross, C. Adamo, J. Jaramillo, R. Gomperts, R. E. Stratmann, O. Yazyev, A. J. Austin, R. Cammi, C. Pomelli, J. W. Ochterski, P. Y. Ayala, K. Morokuma, G. A. Voth, P. Salvador, J. J. Dannenberg, V. G. Zakrzewski, S. Dapprich, A. D. Daniels, M. C. Strain, O. Farkas, D. K. Malick, A. D. Rabuck, K. Raghavachari, J. B. Foresman, J. V. Ortiz, Q. Cui, A. G. Baboul, S. Clifford, J. Cioslowski, B. B. Stefanov, G. Liu, A. Liashenko, P. Piskorz, I. Komaromi, R. L. Martin, D. J. Fox, T. Keith, M. A. Al-Laham, C. Y. Peng, A. Nanayakkara, M. Challacombe, P. M. W. Gill, B. Johnson, W. Chen, M. W. Wong, C. Gonzalez, and J. A. Pople, Gaussian, Inc., Wallingford CT, 2004.
17. C. Müller, Private comm. (2006).
18. K. B. Wiberg, Tetrahedron **24** (1968) 1083.
19. T. Steiner and W. Saenger, Acta Cryst. B **50** (1994) 348.
20. T. Steiner, Angew. Chem. Int. Ed. **41** (2002) 48.
21. G. Herzberg, *Molecular Spectra and Molecular Structure*. New York, Van Nostrand (1950).
22. J. A. Ibers, J.Chem. Phys. **41** (1964) 25.
23. W. S. Benedict, N. Gailar, and E. K. Plyler, J. Chem. Phys. **24** (1956) 1139.
24. I. Olovsson and P. G. Jönsson, In: *The Hydrogen Bond*, pp. 393–456. Eds.: P. Schuster, G. Zundel, C. Sandorfy, Amsterdam, North Holland (1976).
25. W. Joswig, H. Fuess, and G. Ferraris, Acta Cryst. B **38** (1982) 2798.
26. G. Herzberg, *Infrared and Raman Spectra of Polyatomic Molecules*. New York, Van Nostrand (1945).
27. B. P. Stoicheff, Can. J. Phys. **35** (1957) 730.
28. B. P. Stoicheff, Adv. Spectr. **1** (1959) 91.

The whole palette of hydrogen bonds

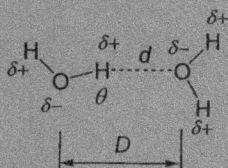

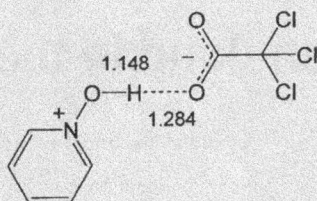

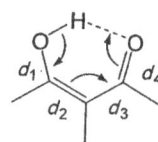

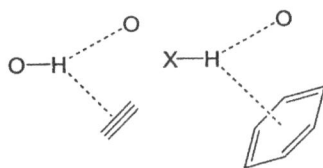

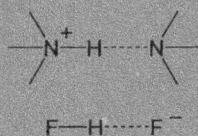

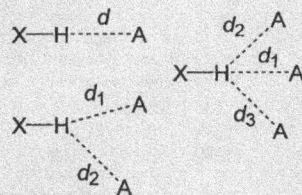

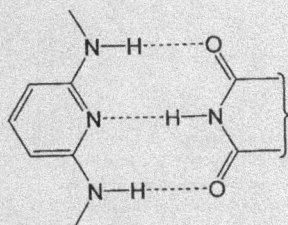

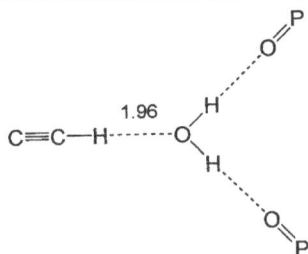

ANGEWANDTE CHEMIE ©WILEY-VCH

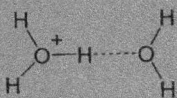

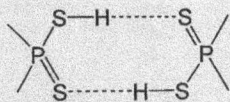

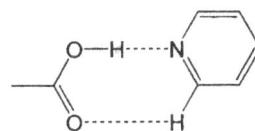

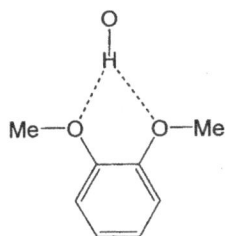

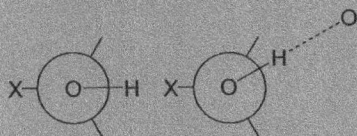

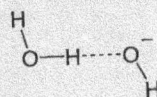

The Hydrogen Bond in the Solid State

Thomas Steiner*

In memory of Jan Kroon

The hydrogen bond is the most important of all directional intermolecular interactions. It is operative in determining molecular conformation, molecular aggregation, and the function of a vast number of chemical systems ranging from inorganic to biological. Research into hydrogen bonds experienced a stagnant period in the 1980s, but re-opened around 1990, and has been in rapid development since then. In terms of modern concepts, the hydrogen bond is understood as a very broad phenomenon, and it is accepted that there are open borders to other effects. There are dozens of different types of X–H ··· A hydrogen bonds that occur commonly in the condensed phases, and in addition there are innumerable less common ones. Dissociation energies span more than two orders of magnitude (about 0.2 – 40 kcal mol^{-1}). Within this range, the nature of the interaction is not constant, but its electrostatic, covalent, and dispersion contributions vary in their relative weights. The hydrogen bond has broad transition regions that merge continuously with the covalent bond, the van der Waals interaction, the ionic interaction, and also the cation – π interaction. All hydrogen bonds can be considered as incipient proton transfer reactions, and for strong hydrogen bonds, this reaction can be in a very advanced state. In this review, a coherent survey is given on all these matters.

Keywords: donor – acceptor systems · electrostatic interactions · hydrogen bonds · noncovalent interactions · proton transfer

1. Introduction

The hydrogen bond was discovered almost 100 years ago,[1] but still is a topic of vital scientific research. The reason for this long-lasting interest lies in the eminent importance of hydrogen bonds for the structure, function, and dynamics of a vast number of chemical systems, which range from inorganic to biological chemistry. The scientific branches involved are very diverse, and one may include mineralogy, material science, general inorganic and organic chemistry, supramolecular chemistry, biochemistry, molecular medicine, and pharmacy. The ongoing developments in all these fields keep research into hydrogen bonds developing in parallel. In recent years in particular, hydrogen-bond research has strongly expanded in depth as well as in breadth, new concepts have been established, and the complexity of the phenomena considered has increased dramatically. This review is intended to give a coherent survey of the state of the art, with a focus on the structure in the solid state, and with weight put mainly on the fundamental aspects. Numerous books[2–9] and reviews on the subject have appeared earlier, so a historical outline is not necessary. Much of the published numerical material is somewhat outdated and, therefore, this review contains some numerical data that have been newly retrieved from the most relevant structural database, the Cambridge Structural Database (CSD).[10]

It is pertinent to recall here the earlier "classical" view on hydrogen bonding. One may consider the directional interaction between water molecules as the prototype of all hydrogen bonds (Scheme 1, definitions of geometric parameters are also included). The large difference in electronegativity between the H and O atoms makes the O–H bonds of a water molecule inherently polar, with partial atomic charges of around +0.4 on each H atom and –0.8 on the O atom. Neighboring water molecules orient in such a way that local dipoles O$^{\delta-}$–H$^{\delta+}$ point at negative partial charges O$^{\delta-}$, that is, at the electron lone pairs of the filled p orbitals. In the resulting

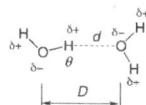

Scheme 1. Prototype of a hydrogen bond: the water dimer. Definitions of geometrical parameters: d = H ··· O distance, D = O ··· O distance, θ = O–H ··· O angle.

[*] Dr. T. Steiner
    Institut für Chemie—Kristallographie
    Freie Universität Berlin
    Takustrasse 6, 14195 Berlin (Germany)
    Fax: (+ 49) 30-838-56702
    E-mail: steiner@chemie.fu-berlin.de

REVIEWS

T. Steiner

O–H$\cdots$|O interaction, the intermolecular distance is shortened by around 1 Å compared to the sum of the van der Waals radii for the H and O atoms[11] (1 Å = 100 pm), which indicates there is substantial overlap of electron orbitals to form a three-center four-electron bond. Despite significant charge transfer in the hydrogen bond, the total interaction is dominantly electrostatic, which leads to pronounced flexibility in the bond length and angle. The dissociation energy is around 3 – 5 kcal mol^{-1}.

This brief outline of the hydrogen bond between water molecules can be extended, with only minor modifications, to analogous interactions X–H$\cdots$A formed by strongly polar groups X$^{\delta-}$–H$^{\delta+}$ on one side, and atoms A$^{\delta-}$ on the other (X = O, N, halogen; A = O, N, S, halide, etc.). Many aspects of hydrogen bonds in structural chemistry and structural biology can be readily explained at this level, and it is certainly the relative success of these views that made them dominate the perception of the hydrogen bond for decades. This dominance has been so strong in some periods that research on hydrogen bonds differing too much from the one between water molecules was effectively impeded.[8]

Today, it is known that the hydrogen bond is a much broader phenomenon than sketched above. What can be called the "classical hydrogen bond" is just one among many—a very abundant and important one, though. We know of hydrogen bonds that are so strong that they resemble covalent bonds in most of their properties, and we know of others that are so weak that they can hardly be distinguished from van der Waals interactions. In fact, the phenomenon has continuous transition regions to such different effects as the covalent bond, the purely ionic, the cation–π, and the van der Waals interaction. The electrostatic dominance of the hydrogen bond is true only for some of the occurring configurations, whereas for others it is not. The H$\cdots$A distance is not in all hydrogen bonds shorter than the sum of the van der Waals radii. For an X–H group to be able to form hydrogen bonds, X does not need to be "very electronegative", it is only necessary that X–H is at least slightly polar. This requirement includes groups such as C–H, P–H, and some metal hydrides. X–H groups of reverse polarity, X$^{\delta+}$–H$^{\delta-}$, can form directional interactions that parallel hydrogen bonds (but one can argue that they should not be called so). Also, the counterpart A does not need to be a

particularly electronegative atom or an anion, but only has to supply a sterically accessible concentration of negative charge. The energy range for dissociation of hydrogen bonds covers more than two factors of ten, about 0.2 to 40 kcal mol^{-1}, and the possible functions of a particular type of hydrogen bond depend on its location on this scale. These issues shall all be discussed in the following sections.

For space reasons, it will not be possible to cover all aspects of hydrogen bonding equally well. Therefore, some important fields, for which recent guiding reviews are available, will not be discussed in great length. One example is the role of hydrogen bonds in molecular recognition patters ("supramolecular synthons"),[12] and the use of suitably robust motifs for the construction of crystalline architectures with desired properties ("crystal engineering").[13, 14] This area includes the interplay of hydrogen bonds with other intermolecular forces, with whole arrays of such forces, and hierarchies within such an interplay. The reader interested in this complex field is referred to the articles of Desiraju,[12, 13] Leiserowitz et al.,[15] and others.[16] A further topic which could not be covered here is the symbolic description of hydrogen bond networks using tools of graph theory,[17] in particular the "graph set analysis".[18] An excellent guiding review is also available in this case.[19] For hydrogen bonding in biological structures, the interested reader is referred to the book of Jeffrey and Saenger,[5] and for theoretical aspects to the book of Scheiner[7] as well as other recent reviews.[20] Results obtained with experimental methods other than diffraction will be touched upon only briefly, and will possibly leave some readers dissatisfied. The role of hydrogen bonding in special systems will not be discussed at all, simply because there are too many of them.

2. Fundamentals

2.1. Definition of the Hydrogen Bond

Before discussing the hydrogen bond itself, the matter of hydrogen bond definitions must be addressed. This is an important point, because definitions of terms often limit entire fields. It is, also, a problematic point because very different hydrogen bond definitions have been made, and part

Thomas Steiner, born in 1961 near Reutte/Tirol in Austria, studied experimental physics at the Technical University Graz, and obtained his Ph.D. in 1990 at the Freie Universität Berlin with Prof. Wolfram Saenger. He completed his habilitation in 1996, also at the Freie Universität Berlin (Faculty of Chemistry). He was guest scientist with Prof. Jan Kroon at the Bijvoet Center for Biomolecular Research at Utrecht University, The Netherlands in 1995, and with Prof. Joel L. Sussman at the Department of Structural Biology at the Weizmann Institute of Science, Israel in 1997/8. His research interests are hydrogen bonds and other intermolecular interactions in structural chemistry and biology. The methods used for investigation are neutron and X-ray diffraction, neutron scattering, crystal engineering, IR spectroscopy, database analysis, and crystal correlation. The systems studied range from organic (terminal alkynes, binary proton transfer complexes) and bioorganic (peptides, steroids) to biological (proteins). Together with Prof. G. R. Desiraju, he is an author of a book on weak hydrogen bonds.

Angew. Chem. Int. Ed. **2002**, *41*, 48 – 76

Snow, Ice and Other Wonders of Water

REVIEWS

of the literature relies quite uncritically on the validity (or the value) of the particular definition that is adhered to.

Time has shown that only very general and flexible definitions of the term "hydrogen bond" can do justice to the complexity and chemical variability of the observed phenomena, and include the strongest as well as the weakest species of the family, and inter- as well as intramolecular interactions. A far-sighted early definition is that of Pimentel and McClellan, who essentially wrote that "...a hydrogen bond exists if 1) there is evidence of a bond, and 2) there is evidence that this bond sterically involves a hydrogen atom already bonded to another atom".[2] This definition leaves the chemical nature of the participants, including their polarities and net charges, unspecified. No restriction is made on the interaction geometry except that the hydrogen atom must be somehow "involved". The crucial requirement is the existence of a "bond", which is itself not easy to define. The methods to test experimentally if requirements 1 and 2 are fulfilled are limited. For crystalline compounds, it is easy to see with diffraction experiments whether an H atom is involved, but it is difficult to guarantee that a given contact is actually "bonding".

A drawback of the Pimentel and McClellan definition is that in the strict sense it includes pure van der Waals contacts (which can be clearly "bonding", with energies of several tenths of a kcal mol^{-1}), and it also includes three-center two-electron interactions where electrons of an X–H bond are donated sideways to an electron-deficient center ("agostic interaction"). From a modern viewpoint, it seems advisable to modify point 2, such as by requiring that X–H acts as a proton (not electron) donor. Therefore, the following definition is proposed:

An X–H ··· A interaction is called a "hydrogen bond", if 1. it constitutes a local bond, and 2. X–H acts as proton donor to A.

The second requirement is related to the acid/base properties of X–H and A, and has the chemical implication that a hydrogen bond can at least in principle be understood as an incipient proton-transfer reaction from X–H to A. It excludes, for example, pure van der Waals contacts, agostic interactions, so-called "inverse hydrogen bonds" (see Section 8), and B-H-B bridges. As a matter of fact, point 2 should be interpreted liberally enough to include symmetric hydrogen bonds X–H–X, where donor and acceptor cannot be distinguished. The direction of formal or real electron transfer in a hydrogen bond is reverse to the direction of proton donation.

Apart from general chemical definitions, there are many specialized definitions of hydrogen bonds that are based on certain sets of properties that can be studied with a particular technique. For example, hydrogen bonds have been defined on the basis of interaction geometries in crystal structures (short distances, fairly "linear angles" θ), certain effects in IR absorption spectra (red-shift and intensification of ν_{XH}, etc.), or certain properties of experimental electron density distributions (existence of a "bond critical point" between H and A, with numerical parameters within certain ranges). All such definitions are closely tied to a specific technique, and may be

quite useful in the regime accessible to it. Nevertheless, they are more or less useless outside that regime, and many a misunderstanding in the hydrogen bond literature has been caused by applying such definitions outside their region of applicability.

The practical scientist often requires a technical definition, and automated data treatment procedures for identifying hydrogen bonds cannot be done without. It is not within the scope of this article to discuss any set of threshold values that a "hydrogen bond" must pass in any particular type of technical definition. It is only mentioned that the "van der Waals cutoff" definition[21] for identifying hydrogen bonds on a structural basis (requiring that the H ··· A distance is substantially shorter than the sum of the van der Waals radii of H and A) is far too restrictive and should no longer be applied.[5, 6, 8] If distance cutoff limits must be used, X–H ··· A interactions with H ··· A distances up to 3.0 or even 3.2 Å should be considered as potentially hydrogen bonding.[6] An angular cutoff can be set at $>90°$ or, somewhat more conservatively, at $>110°$. A necessary geometric criterion for hydrogen bonding is a positive directionality preference, that is, linear X–H ··· A angles must be statistically favored over bent ones (this is a consequence of point 2 of the above definition).[22]

2.2. Further Terminology

A large part of the terminology concerning hydrogen bonds is not uniformly used in the literature, and still today, terminological discrepancies lead to misunderstanding between different authors. Therefore, some of the technical terms used in this review need to be explicitly defined.

In a hydrogen bond X–H ··· A, the group X–H is called the *donor* and A is called the *acceptor* (short for "proton donor" and "proton acceptor", respectively). Some authors prefer the reverse nomenclature (X–H = electron *acceptor*, Y = electron *donor*), which is equally justified.

In a simple hydrogen bond, the donor interacts with one acceptor (Scheme 2a). Since the hydrogen bond has a long range, a donor can interact with two and three acceptors simultaneously (Scheme 2b, c). Hydrogen bonds with more than three acceptors are possible in principle, but are only rarely found in practice because they require very high spatial densities of acceptors. The terms "bifurcated" and "trifurcated" are commonly used to describe the arrangements in Scheme 2b and c, respectively. The term "two-centered" hydrogen bond is an alternative descriptor for X–H ··· A (Scheme 2a) where the H-atom is bonded to *two* other atoms, and is itself not

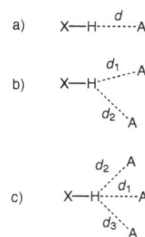

Scheme 2. Different types of hydrogen bridges. a) Normal hydrogen bond with one acceptor. b) Bifurcated hydrogen bond; if the two H ··· A separations are distinctly different, the shorter interaction is called major component, and the longer one the minor component of the bifurcated bond. c) Trifurcated hydrogen bond.

REVIEWS

counted as a center. Consequently, the arrangements in Scheme 2b and 2c may be called "three-" and "four-centered" hydrogen bonds, respectively.[5, 6] This terminology is logical, but leads to confusion from the point of view of regarding hydrogen bonds O–H ⋯ O as "three-center four-electron" interactions, where the H-atom *is* counted as a center. A bifurcated hydrogen bond (Scheme 2b) is then termed "three-centered", but also represents a "four-center six-electron" interaction. To avoid such ambiguities, the older term "bifurcated" is used here.

There is particular confusion concerning the terms *attractive* and *repulsive*. Some authors use these terms to characterize forces, and others to characterize energies. In the latter case, an "attractive interaction" is taken as a synonym for "bonding interaction", that is, one that requires the input of energy to be broken. Following well-founded recommendations,[23] the terms "attractive" and "repulsive" are used here exclusively to describe forces. Negative and positive bond energies are indicated by the terms "stabilizing" (or "bonding") and "destabilizing", respectively. The schematic hydrogen bond potential in Figure 1 shows that a stabilizing interaction (that is, with $E < 0$) is associated with a repulsive force if it is shorter than the equilibrium distance (see figure legend for further details).[8]

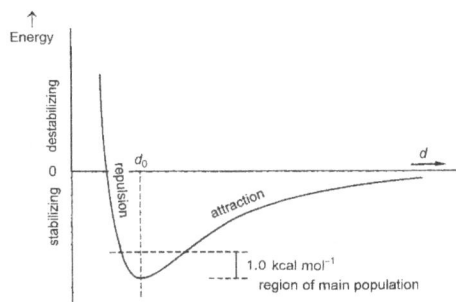

Figure 1. Schematic representation of a typical hydrogen bond potential.[8] A hydrogen bond length differing from d_0 implies a force towards a geometry of lower energy, that is, by attraction if $d > d_0$ and repulsion if $d < d_0$. Note that the interaction can at the same time be "stabilizing" (or "bonding") and "repulsive"! The distortions from d_0 occurring in practice are limited by the energy penalties that have to be paid, and in crystals, only a few hydrogen bonds have energies differing by more than 1 kcal mol^{-1} from optimum.

Hydrogen bonds are sometimes called "nonbonded interactions". At least to this author, this appears a contradiction in terms which should be avoided.

2.3. Constituent Interactions

The hydrogen bond is a complex interaction composed of several constituents that are different in their natures.[6, 7] Most popular are partitioning modes that essentially follow those used by Morokuma.[24] The total energy of a hydrogen bond (E_{tot}) is split into contributions from electrostatics (E_{el}),

polarization (E_{pol}), charge transfer (E_{ct}), dispersion (E_{disp}), and exchange repulsion (E_{er}), somewhat different, but still related, partitioning schemes are also in use. The distance and angular characteristics of these constituents are very different. The electrostatic term is directional and of long range (diminishing only slowly as $-r^{-3}$ for dipole–dipole and as $-r^{-2}$ for dipole–monopole interactions). Polarization decreases faster ($-r^{-4}$) and the charge-transfer term decreases even faster, approximately following e^{-r}. According to natural bond orbital analysis,[25] charge transfer occurs from an electron lone pair of A to an antibonding orbital of X–H, that is $n_A \rightarrow \sigma^*_{XH}$. The dispersion term is isotropic with a distance dependence of $-r^{-6}$. The exchange repulsion term increases sharply with reducing distance (as $+r^{-12}$). The dispersion and exchange repulsion terms are often combined into an isotropic "van der Waals" contribution that is approximately described by the well-known Lennard–Jones potential ($E_{vdW} \sim A r^{-12} - B r^{-6}$). Depending on the particular chemical donor–acceptor combination, and the details of the contact geometry, all these terms contribute with different weights. It cannot be globally stated that the hydrogen bond as such is dominated by this or that term in any case.

Some general conclusions can be drawn from the overall distance characteristics. In particular, it is important that of all the constituents, the electrostatic contribution reduces slowest with increasing distance. The hydrogen bond potential for any particular donor–acceptor combination (Figure 1) is, therefore, dominated by electrostatics at long distances, even if charge transfer plays an important role at optimal geometry. Elongation of a hydrogen bond from optimal geometry *always* makes it more electrostatic in nature.

In "normal" hydrogen bonds E_{el} is the largest term, but a certain charge-transfer contribution is also present. The van der Waals terms too are always present, and for the weakest kinds of hydrogen bonds dispersion may contribute as much as electrostatics to the total bond energy. Purely "electrostatic plus van der Waals" models can be quite successful despite their simplicity for hydrogen bonds of weak to intermediate strengths.[26] Such simple models fail for the strongest types of hydrogen bonds, for which their quasi-covalent nature has to be fully considered (see Section 7).

2.4. Energies

The energy of hydrogen bonds in the solid state cannot be directly measured, and this circumstance leaves open questions in many structural studies. Computational chemistry, on the other hand, produces results on hydrogen bond energies at an inflationary rate,[7, 20] many obtained at high levels of theory and even more in rather routine calculations using black-box methods. Theoretical studies are not the topic of the present review, but an idea of typical results can be gained from the collection of calculated values listed in Table 1.[27] It appears that hydrogen bond energies cover more than two orders of magnitude, about -0.2 to -40 kcal mol^{-1}. On a logarithmic scale, the bond energy of the water dimer is roughly in the middle.

Solid-State Hydrogen Bonds   **REVIEWS**

Table 1. Calculated hydrogen bond energies (kcal mol^{-1}) in some gas-phase dimers.[a]

Dimer	Energy	Ref.
[F–H–F]$^-$	39	[27a]
[H$_2$O–H–OH$_2$]$^+$	33	[27b]
[H$_3$N–H–NH$_3$]$^+$	24	[27b]
[HO–H–OH]$^-$	23	[27a]
NH$_4^+$ $\cdots$ OH$_2$	19	[27c]
NH$_4^+$ $\cdots$ Bz	17	[27d]
HOH $\cdots$ Cl$^-$	13.5	[27c]
O=C–OH $\cdots$ O=C–OH	7.4	[27e]
HOH $\cdots$ OH$_2$	4.7; 5.0	[27f,g]
N≡C–H $\cdots$ OH$_2$	3.8	[27h]
HOH $\cdots$ Bz	3.2	[27i]
F$_3$C–H $\cdots$ OH$_2$	3.1	[27j]
Me–OH $\cdots$ Bz	2.8	[27k]
F$_2$HC–H $\cdots$ OH$_2$	2.1; 2.5	[27f,j]
NH$_3$ $\cdots$ Bz	2.2	[27i]
HC≡CH $\cdots$ OH$_2$	2.2	[27h]
CH$_4$ $\cdots$ Bz	1.4	[27i]
FH$_2$C–H $\cdots$ OH$_2$	1.3	[27f,j]
HC≡CH $\cdots$ C≡CH$^-$	1.2	[27l]
HSH $\cdots$ SH$_2$	1.1	[27m]
H$_2$C=CH$_2$ $\cdots$ OH$_2$	1.0	[27l]
CH$_4$ $\cdots$ OH$_2$	0.3; 0.5; 0.6; 0.8	[27f,n–p]
C=CH$_2$ $\cdots$ C=C	0.5	[27l]
CH$_4$ $\cdots$ F–CH$_3$	0.2	[27q]

[a] For computational details, see the original literature. Bz = benzyl.

The values in Table 1 are computed for dimers in optimal geometry undisturbed by their surroundings. In the solid state, hydrogen bonds are practically never in optimal geometry, and are always influenced by their environment. There are numerous effects from the close and also from the remote surrounding that may considerably increase or lower hydrogen bond energies ("crystal-field effects"). Hydrogen bonds do not normally occur as isolated entities but form networks. Within these networks, hydrogen bond energies are not additive (see Section 4). In such cases, it is not reasonable to split up the network into individual hydrogen bonds and to calculate energies for each one. In this sense, calculated hydrogen bond energies should always be taken with caution.

2.5. Transition to Other Interaction Types

As outlined previously, the hydrogen bond is composed of several constituent interactions which are variant in their contributing weights. Chemical variation of donor and/or acceptor, and possibly also of the environment, can gradually change a hydrogen bond to another interaction type. This shall be detailed here for the most important cases.

The transition to pure van der Waals interaction is very common. The polarity of X–H or A (or both) in the array X$^{\delta-}$–H$^{\delta+}$ $\cdots$ A$^{\delta-}$ can be reduced by suitable variation of X or A. This reduces the electrostatic part of the interaction, whereas the van der Waals component is much less affected. In consequence, the van der Waals component gains relative weight, and the angular characteristics gradually change from directional to isotropic. Since the polarities of X$^{\delta-}$–H$^{\delta+}$ or A$^{\delta-}$ can be reduced to zero continuously, the resulting transition of the interaction from hydrogen bond to van der Waals type is

continuous too. Such a behavior was actually demonstrated for the directionality of C–H $\cdots$ O=C interactions, which gradually disappears when the donor is varied from C≡C–H to C=CH$_2$ to C–CH$_3$ (see Figure 8, Section 3.2).[22] At the acceptor side of a hydrogen bond, sulfur is typical of an atom that allows continuous variation of the partial charge from S$^{\delta-}$ to S$^{\delta+}$. Therefore, one can create a continuum of chemical situations between the S atom acting as a fairly strong hydrogen bond acceptor, and being inert to hydrogen bonding (the extreme cases are ionic species such as X–S$^-$ and X=S$^+$–Y).

At the other end of the energy scale, there is a continuous transition to covalent bonding.[28] In the so-called symmetric hydrogen bonds X–H–X, where an H atom is equally shared between two chemically identical atoms X, no distinction can be made between a donor and an acceptor, or a "covalent" X–H and "noncovalent" H $\cdots$ X bond (found experimentally for X = F, O, and possibly N). In fact, this situation can be conveniently described as a hydrogen atom forming two covalent bonds with bond orders $s = \frac{1}{2}$. In crystals (and also in solution), all intermediate cases exist between the extremes X–H $\cdots\cdots$ IX and X–H–X. Strongly covalent hydrogen bonds will be discussed in greater detail in Section 7, and the bond orders ("valences") of H $\cdots$ O over the whole distance range will be given in Section 9 (Table 7).

There is also a gradual transition from hydrogen bonding to purely ionic interactions. If in an interaction X$^{\delta-}$–H$^{\delta+}$ $\cdots$ Y$^{\delta-}$–H$^{\delta+}$ the net charges on X–H and Y–H are zero, the electrostatics are of the dipole–dipole type. In general, however, the net charges are not zero. Alcoholic O–H groups have a partial negative charge in addition to their dipole moment, ammonium groups have a positive net charge, and so on. This situation leads to ionic interactions between the charge centers with the energy having a r^{-1} distance dependence. If the charges are large, the ionic behavior may become dominant. For fully charged hydrogen bond partners, energetics are typically dominated by the Coulombic interaction between the charge centers, but the total interaction still remains directional, with X–H not oriented at random but pointing at A. An important example are the so-called salt-bridges between primary ammonium and carboxylate groups in biological structures,[5] N$^+$–H $\cdots$ O$^-$. If weakly polar X–H groups are attached to a charged atom, such as the methyl groups in the N$^+$Me$_4$ ion, they are often involved in short contacts to an approaching counterion, N$^+$–X–H $\cdots$ A$^-$.[8] Although these interactions are directional and may still be classified as a kind of hydrogen bond, their dominant part is certainly the ionic bond N$^+$ $\cdots$ A$^-$.

Finally, there is a transition region between the hydrogen bond and the cation–π interaction. In the pure cation–π interaction a spherical cation such as K$^+$ contacts the negative charge concentration of a π-bonded moiety such as a phenyl ring. This can be considered an electrostatic monopole–quadrupole interaction. The bond energy is -19.2 kcal mol^{-1} for the example of K$^+$ $\cdots$ benzene.[29] A pure π-type hydrogen bond X$^{\delta-}$–H$^{\delta+}$ $\cdots$ Ph is formally a dipole–quadrupole interaction with much lower energies of only a few kcal mol^{-1} (Table 1). If charged hydrogen bond donors such as NH$_4^+$ interact with π-electron clouds, local dipoles are oriented at the π face,[30] but the energetics are dominated by the charge – quadrupole interaction[27d] ($NH_4^+ \cdots Bz$ experimentally: -19.3 kcal mol^{-1}).[29] If the X–H groups of the cation are only weakly polar, they may also orient at the π face and cause some modulation of the dominant cation – π interaction, but this modulation fades to zero with decreasing X–H polarity.

2.6. Incipient Proton Transfer Reaction

A very important way of looking at hydrogen bonds is to regard them as incipient proton-transfer reactions. From this viewpoint, a stable hydrogen bond X–H $\cdots$ Y is a "frozen" stage of the reaction X–H $\cdots$ Y $\rightleftarrows$ X$^-$ $\cdots$ H$^-$$^+$Y (or X$^+$–H $\cdots$ Y $\rightleftarrows$ X $\cdots$ H$^-$$^+$Y, etc.). This means that a partial bond H $\cdots$ Y is already established and the X–H bond is concomitantly weakened.[31] In the case of strong hydrogen bonds, the stage of proton transfer can be quite advanced. In some hydrogen bonds the proton position is not stable at X or Y, but proton transfer actually takes place with high rates. In other cases these rates are small or negligible.

The interpretation of hydrogen bonds as an incipient chemical reaction is complementary to electrostatic views on hydrogen bonding. It brings into play acid – base considerations, proton affinities, the partially covalent nature of the H $\cdots$ Y bond, and turns out to be a very powerful concept for understanding the stronger types of hydrogen bonds in particular. For example, the partial H $\cdots$ Y bond can only become strong if its orientation roughly coincides with the orientation of the full H–Y bond that would be formed upon proton transfer. Approach in different orientations may still be favorable in electrostatic terms, but results only in moderately strong hydrogen bonds.

This view also helps in deciding whether a particular type of X–H $\cdots$ A interaction may be classified as a hydrogen bond or not (compare the definition in Section 2.1). *Only* if it may be thought of as a frozen proton-transfer reaction, may it be called a hydrogen bond.

2.7. Location of the H Atom

An atom is constituted of a nucleus and its electron shell. Normally, the centers of gravity of the nucleus and electron shell coincide well, and this common center is called the "location" of the atom. For H atoms, however, this is generally not the case. In a covalent bond with a more electronegative atom, the average position of the single electron of the H atom is displaced towards that other atom. The centers of gravity of the nucleus and electron no longer coincide, and this leads to a conceptual problem: what should be taken as the "location" of the atom? It is not chemically reasonable to consider one of the two centers of gravity as the "right" location of the atom, and the other as "wrong", but one must accept that a point-atom model is simplistic in this situation.[32, 33] In practice, this leads to unpleasant complications. X-ray diffraction experiments determine electron-density distributions and locate the electron-density maxima of the atoms. Neutron diffraction, on the other hand, locates

the nuclei. The results of the two techniques for H atoms often differ by more than 0.1 Å.[34] Neither of the two results is more true than the other, but they are complementary and both represent useful pieces of information. Nevertheless, neutron diffraction results are much more precise and reliable, and allow the proton positions to be located as accurately as other nuclei.

It has become a practice in the analysis of X-ray diffraction results to "normalize" the X–H bonds by shifting the position found for the H atom (that is, the position of the electron center of gravity) along the X–H vector to the average neutron-determined internuclear distance, namely, to the approximate position of the proton.[35] This theoretical position is then used for the calculation of hydrogen bond parameters. The currently used standard bond lengths are: O–H = 0.983, N–H = 1.009, C–H = 1.083, B–H = 1.19, and S–H = 1.34 Å; a more complete list can be found in ref. [8]. The normalization procedure is generally reasonable, well suited to smooth out the large experimental uncertainty of X-ray diffraction data, and is particularly useful in statistical database analysis. Nevertheless, one must be aware that it is not a correction in the strict sense, instead it replaces a certain structural feature (the location of the electron center of gravity) by a chemically different one (the proton position). Furthermore, the internuclear X–H bond length is fairly constant only in weak and moderate hydrogen bonds, whereas it is significantly elongated in strong ones. In the latter situation, the elongation should at least in principle be taken into account in the normalization. This requires, however, knowledge of the relationship between the relevant X–H and H $\cdots$ A distances (see Section 3.6).[36]

2.8. Charge Density Properties

The precise mapping of the distribution of charge density in hydrogen-bonded systems is a classical topic in structural chemistry,[37] with a large number of individual studies reported.[38] Currently, Baders quantum theory of atoms in molecules (AIM) is the most frequently used formalism in theoretical analyses of charge density.[39] Each point in space is characterized by a charge density $\rho(r)$, and further quantities such as the gradient of $\rho(r)$, the Laplacian function of $\rho(r)$, and the matrix of the second derivatives of $\rho(r)$ (Hessian matrix). The relevant definitions and the topology of $\rho(r)$ in a molecule or molecular complex can be best understood with the help of an illustration (Figure 2; see figure legend for details).[40] The thin lines represent lines of steepest ascent through $\rho(r)$ (trajectories). If there is a chemical bond between two atoms (such as a hydrogen bond), they are directly connected by a trajectory called the "bond path". The point with the minimal ρ value along the bond path is called the "bond critical point" (BCP). It represents a saddle point of $\rho(r)$ (strictly speaking, trajectories terminate at the BCP, so that the bond path represents a pair of trajectories each of which connects a nucleus with the BCP). Different kinds of chemical bonds have different numerical properties at the BCP, such as different electron density ρ_{BCP} and different

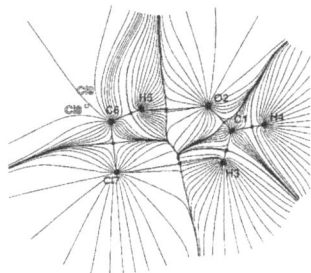

Figure 2. Representative topology of a theoretical electron density function in a hydrogen-bonded system: the adduct of chloroform and formaldehyde formed through a C–H ⋯ O hydrogen bond.[40] Thin lines represent lines of steepest ascent through $\rho(r)$ (trajectories). Critical points (CP) of $\rho(r)$ are maxima and points where the first derivative vanishes. There are four types of CPs in three-dimensional space (rank 3, that is, nondegenerate). Maxima are denoted (3, −3) and minima (3, +3). The latter are also called "cage critical points" (CCP). Saddle points representing a minimum in one direction of space and maxima in two perpendicular directions are called "bond critical points" (BCP) and denoted (3, −1). Saddle points, which represent minima in two perpendicular directions of space and a maximum in the third direction, are called "ring critical points" (RCP) and denoted (3, +1). The trajectories ending at a nucleus constitute a "molecular basin". Basins of neighboring atoms are separated by trajectories that do not end at nuclei (namely, the "interatomic surface"). Trajectories connecting nuclei through a BCP are called a bond path. The electron density at BCPs are minima in the bond paths and maxima in the interatomic surface. BCPs are shown in the figure as squares.

values of the Laplacian function (negative for covalent bonds and H ⋯ A interactions of very strong hydrogen bonds, and positive for ionic bonds, van der Waals interactions, H ⋯ A interactions of medium strength, and weak hydrogen bonds).

The electron density at the bond critical point (ρ_{BCP}) is higher in strong bonds than in weak ones. There are two bond critical points in a hydrogen bond X–H ⋯ A, one between X and H, and one between H and A. In normal hydrogen bonds, the ρ_{BCP} value in X–H is much larger than in H ⋯ A. The value of ρ_{BCP} in H ⋯ A increases with increasing hydrogen bond strength, while that in X–H decreases concomitantly. In the ideally centered case, X–H–X, ρ_{BCP} is equal for both bonds. This behavior has been nicely illustrated for O–H ⋯ O hydrogen bonds (Figure 3).[41] Bond paths with

significant values of ρ_{BCP} have been calculated also for weaker hydrogen bonds of the types C–H ⋯ O[40] and C–H ⋯ π[27l], as well as for "dihydrogen bonds".[42] Electron-density properties of the agostic interaction relative to the hydrogen bond have also been characterized.[43]

Hydrogen bond properties are sometimes discussed exclusively in terms of topological analysis of theoretical $\rho(r)$ distributions. Despite the merits of the method, it is unfortunate that discussion tends to be very formalistic, occasionally even overriding conflicts with experimental data.

2.9. IR and NMR Spectroscopic Properties

IR and NMR spectroscopy have both become standard methods to investigate hydrogen bonds in the solid state.[6] Nevertheless, they are not the focus of the present article and are, therefore, only briefly touched on here.

Formation of a hydrogen bond affects the vibrational modes of the groups involved in several ways.[44] For relatively simple systems, these effects can be studied quantitatively by solid-state IR spectroscopy. If there are many symmetry-independent bonded groups, however, band overlap normally prevents detailed analysis. The frequency of the donor X–H stretching vibration ($\tilde{v}_{X-H}$) is best studied because it is (for polar X–H groups) quite easy to identify in absorption spectra, and in most cases very sensitive to the formation of hydrogen bonds (red-shift of the absorption band, band broadening or intensification). For O–H ⋯ O hydrogen bonds, $\tilde{v}_{O-H}$ is correlated with the O ⋯ O distance (Figure 4).[45, 46] Analogous correlations have been established

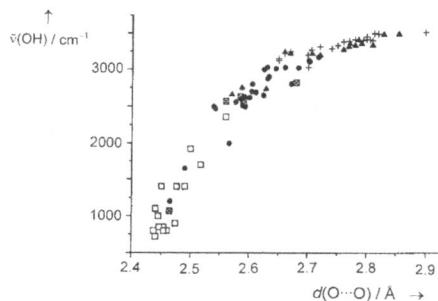

Figure 4. Scatter plot of IR stretching frequencies $\tilde{v}_{OH}$ against O ⋯ O distances in O–H ⋯ O hydrogen bonds (squares: combination of acid and complementary base; filled circles: resonance-assisted hydrogen bonds (RAHB); triangles: σ-cooperative hydrogen bonds; crosses: isolated hydrogen bonds).[46] (IR data from ref. [45a].)

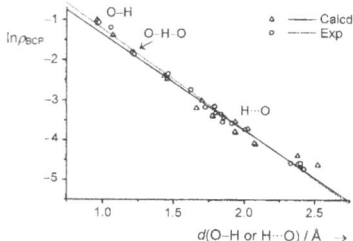

Figure 3. Electron density at the bond critical points, ρ_{BCP}, for a set of O–H ⋯ O hydrogen bonds, together with fitted logarithmic relationships for experimental and theoretical data. (Adapted from ref. [41].)

also for less common hydrogen bond types, for example, between the donor C≡C–H and the acceptors O,[47] N,[48] and C=C.[49] There is considerable scatter in these correlations, not just because of experimental inaccuracy. The correlations represent systematic trends between different physical quantities, but not strict laws (for the discussion of an example, see footnote [50]).

The difference between the $\tilde{v}_{X-H}$ value of free and hydrogen-bonded X–H groups, $\Delta\tilde{v}_{X-H}$, increases systematically with decreasing H $\cdots$ A (or X $\cdots$ A) distance. It has even been reported that a common correlation $\Delta\tilde{v}_{X-H} = f(H \cdots A)$ is approximately valid for many different types of X–H $\cdots$ A hydrogen bonds, and on the basis of a set of diverse organic and inorganic structures, it has been parametrized as $\Delta\tilde{v}_{X-H} = 0.011\, d_{HA}^{-6.1}$ ($\tilde{v}_{X-H}$ in cm^{-1}, d in nm).[51] An approximate relationship with bond enthalpies has also been established for O–H $\cdots$ O hydrogen bonds, $-\Delta H = 0.134\, d_{HA}^{-3.05}$ (H in kJ mol^{-1}, d in nm; the equivalent relationship with band shifts is $-\Delta H = 1.3\,(\Delta\tilde{v})^{0.5}$).[51] The predictive power of these correlations is limited by the large scatter.

Further important properties of v_{X-H} are the band width and the integrated band intensity $I(\tilde{v}_{X-H})$. The band width and $I(\tilde{v}_{X-H})$ increase strongly upon formation of a hydrogen bond, and this is often taken as a more reliable indicator of hydrogen bond formation than the red-shift of v_{X-H}. For example, there are cases of C–H $\cdots$ O hydrogen bonding where Δv_{X-H} is difficult to measure while the increase of $I(\tilde{v}_{X-H})$ is easy to detect.[52] The increase of $I(\tilde{v}_{X-H})$ has been correlated with the strength of the hydrogen bond, and the approximate relationship $-\Delta H = 12.2\,\Delta I(\tilde{v}_{X-H})^{0.5}$ was suggested.[53]

In principle, the H $\cdots$ A stretching vibration is the most direct spectroscopic indicator of hydrogen bonding. For weaker kinds of hydrogen bonds, these bands are in the far infrared, and are investigated only rarely. A direct effect of the hydrogen bond can often be observed also on the acceptor side. In X–H $\cdots$ O=C bonds, for example, the O=C bond is weakened leading to a lowering of the stretching vibration frequency.

The effects described above show many anomalies. For example, bond energies and dissociation constants of C–H $\cdots$ O interactions of chloroform molecules are substantial, but v_{CH} may not only shift to lower, but also to slightly higher wavenumbers. The band intensity always increases, as usual.[54] This effect, long regarded only as an exotic anomaly, has recently attracted greater attention. According to theoretical calculations,[55] a blue-shift of v_{CH} indicates a different kind of electronic interaction in the hydrogen bond: electron density of the acceptor is not mainly transferred into the antibonding σ^* orbital of the donor X–H, but into remote parts of the donor molecule (such as the C–Cl part of CHCl$_3$). This transfer of electron density is also associated with a shortening of the X–H bond. The term "improper blue-shifting" hydrogen bonds was introduced to distinguish these interactions from "proper" hydrogen bonds.

In most hydrogen bonds several nuclei may be observed by NMR spectroscopy. In particular, the proton is increasingly deshielded with increasing hydrogen bond strength, which leads to ^{1}H downfield shifts that are correlated with the length of the hydrogen bond.[6, 56] Thus, NMR shift data can be used to estimate lengths of hydrogen bonds (Figure 5). Chemical shifts of X and A (for example, ^{15}N), X/H and X/A coupling constants, and differences in the ^{1}H and ^{2}H signals in H/D exchange experiments can give additional information on X–H $\cdots$ A bonds. In O=C–OH $\cdots$ N(Py) hydrogen bonds, for example, the ^{15}N chemical shift has been used to probe the protonation state of the N atom: in moderate strength

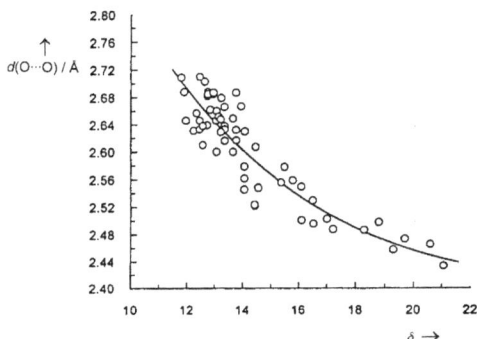

Figure 5. Typical correlation of ^{1}H NMR chemical shift and O $\cdots$ O distance in O–H $\cdots$ O hydrogen bonds.[56c] Other authors have obtained similar figures from different structure samples.[6, 56a,b]

O–H $\cdots$ N hydrogen bonds the shift is <20, in symmetric bonds O–H–N it is around $\delta = -60$, and in ionic bonds O$^-\cdots$ H–N$^+$ it becomes $\delta = -100$.[57] The time scale of proton dynamics in a disordered hydrogen bond can be determined from NMR experiments in a certain frame (μs scale) accessible to the experimental method.

2.10. The Categories of "Strong", "Moderate", and "Weak" Hydrogen Bonds

As we have seen, hydrogen bonds exist with a continuum of strengths. Nevertheless, it can be useful for practical reasons to introduce a classification, such as "weak", "strong", and possibly also "in between". In this article, the system described by Jeffrey is followed,[6] who called hydrogen bonds *moderate* if they resemble those between water molecules or in carbohydrates (one could also call them "normal"), and are associated with energies in the range 4 – 15 kcal mol^{-1}. Hydrogen bonds with energies above and below this range are termed *strong* and *weak*, respectively. Some general properties of these categories are listed in Table 2. It must be stressed that there are no "natural" borderlines between these

Table 2. Strong, moderate, and weak hydrogen bonds following the classification of Jeffrey.[6] The numerical data are guiding values only.

	Strong	Moderate	Weak
interaction type	strongly covalent	mostly electrostatic	electrostat./ dispers.
bond lengths [Å]			
H $\cdots$ A	1.2 – 1.5	1.5 – 2.2	> 2.2
lengthening of X–H [Å]	0.08 – 0.25	0.02 – 0.08	< 0.02
X–H versus H $\cdots$ A	X–H $\approx$ H $\cdots$ A	X–H < H $\cdots$ A	X–H $\ll$ H $\cdots$ A
X $\cdots$ A [Å]	2.2 – 2.5	2.5 – 3.2	> 3.2
directionality	strong	moderate	weak
bond angles [°]	170 – 180	> 130	> 90
bond energy [kcal mol^{-1}]	15 – 40	4 – 15	< 4
relat. IR shift $\Delta\tilde{v}_{XH}$ [cm^{-1}]	25 %	10 – 25 %	< 10 %
^{1}H downfield shift	14 – 22	< 14	

categories, and that there is no point in using this or any related system in too stringent a way. For a comment on the names of the categories, see footnote [58].

3. Geometry

Hydrogen bonds and their environment have a well-defined geometry in the crystalline state. More than 200 000 published organic and organometallic crystal structures provide a vast amount of experimental data that allows hydrogen bond geometries to be analyzed at a high statistical level.[10] The results of such analyses are presented in this section.

3.1. Donor Directionality

The main structural feature distinguishing the hydrogen bond from the van der Waals interaction is preference for linearity. As a typical example, the distribution of angles θ in carbohydrates is shown in Figure 6 (H $\cdots$ O < 2.0 Å). The absolute frequencies peak between 160 and 170° (Figure 6a); these are the angles that occur most frequently in crystals. To obtain the more relevant frequencies per solid angle, one must weight the absolute values with $1/\sin\theta$ ("cone correction").[59] The weighted frequencies clearly peak at linear angles (Figure 6b).

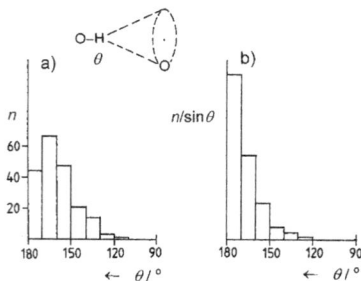

Figure 6. Directionality of O–H $\cdots$ O hydrogen bonds in carbohydrates (H $\cdots$ O < 2.0 Å). a) Conventional histogram with a maximum at slightly bent angles θ. b) Histogram after "cone correction" (weighting with $1/\sin\theta$) which represents the frequency of H-bonds per solid angle.[59] Angular preferences can be seen properly only after cone correction.

The histograms in Figure 6 do not contain distance information, and this is a significant disadvantage. A better (but more costly) way to analyze angular preferences is in scatter plots of angles θ against distances d. This is illutrated for the example of X–H $\cdots$ Cl$^-$ interactions in Figure 7 (for hydroxy donors in Figure 7a and for NH$_3^+$ donors in Figure 7b).[60] The plots include all contacts found in crystal structures with $d < 4.0$ Å at any occurring angle. There are densly populated clusters of data points at short distances and fairly linear angles, and each point in these clusters represents a hydrogen bond. The scatter within the clusters is considerable, and their borders are diffuse. The shortest distances occur at relatively

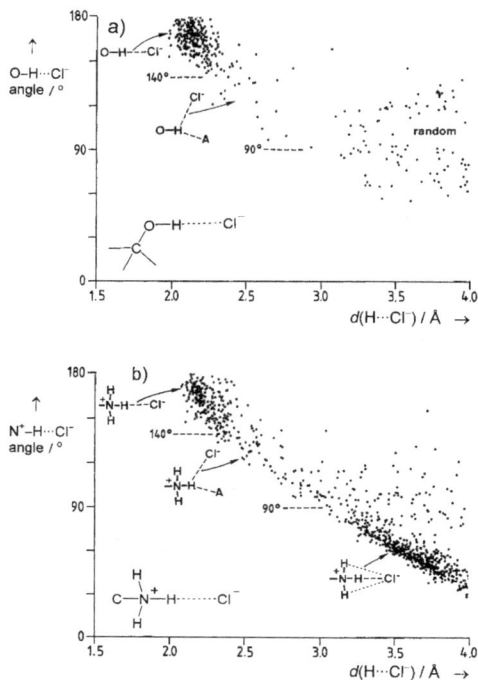

Figure 7. Angular scatter plot of X–H $\cdots$ Cl$^-$ angles against H $\cdots$ Cl$^-$ distances for a) hydroxy and b) –NH$_3^+$ donors (X–H bonds normalized).[60] All contacts with H $\cdots$ Cl$^-$ < 4.0 Å are included, whether they represent a hydrogen bond or not.

linear angles θ, whereas longer bonds are observed with a larger angular range. At longer distances and bent angles, a weakly populated region represents minor components of bifurcated hydrogen bonds (Scheme 2). The regions to the right of these clusters are almost empty, which shows that very long but linear hydrogen bonds almost do not occur (compare with Figure 1). At long distances there is a region of random scatter, which corresponds to X–H groups and chloride ions that do not form a direct interaction. The plots are unpopulated at short distances, because exchange repulsion prevents shorter approach.

The detailed appearance of the plots depends on the type of donor. With the hydroxy group as a donor (Figure 7a), the picture contains only the arrays mentioned above, and the hydrogen bond region is fairly well separated from the region of random scatter. With the more complicated donor –NH$_3^+$ (Figure 7b), there is an additional, densely populated feature at long distances and very bent angles $\theta < 90°$. This new cluster represents the two H atoms of –NH$_3^+$ that point away from the chloride ion if a –NH$_3^+$ $\cdots$ Cl$^-$ hydrogen bond is formed. There is a much higher density of bifurcated hydrogen bonds in Figure 7b than in Figure 7a, and all the populated regions merge into each other. Plots analogous to Figure 7 have been published for other special kinds of

hydrogen bonds, such as O–H···O[61] and C–H···O interactions in carbohydrates,[62] and with more-limited angular
ranges for general O–H···O and N–H···O,[63] water–water,[64] N/O–H···Ph,[65] C–H···Cl(C),[66] and even C–H···
F(C)[67] hydrogen bonds. These figures all show the same
general features (preference of linearity) with some variation
in the details, which indicates that the angular characteristics
of all kinds of hydrogen bonds are related.

The degree of directionality depends on the polarity of the
donor. This effect is shown in Figure 8 with cone-corrected
angular histograms of normal hydrogen bonds O–H···O=C,
of C–H···O=C interactions with three C–H types of different
polarities, and, for comparison, also of C–H···H–C van der
Waals contacts.[22] The degree of directionality decreases in
parallel with the polarity of the X–H group, namely, O–H >
C≡C–H > C=CH₂ > –CH₃. Note that C–H···O contacts of
methyl groups still show a weak but significant preference for
linearity that is clearly different from van der Waals contacts.
This is experimental evidence that methyl C–H···O interactions deserve to be classified as hydrogen bonds (very weak
ones, though).

transfer reaction. The directionality of moderate and weak
hydrogen bonds is much softer, but can still be identified with
the orientation of electron lone pairs (in rare cases[68] also with
filled d_{z^2} orbitals of transition metal atoms). In the classical
example of carbonyl groups, the oxygen lone pair lobes are in
the R_2C=O plane and form angles of about 120° with the C=O
bond. As is seen in an angular histogram with N/O–H donors
(Figure 9, bottom), a corresponding acceptor directionality is
indeed present, but it is softer than is often assumed.[69] A
similar distribution has also been found with stronger types of
C–H donors (C≡CH, Cl₃CH, Cl₂CH₂).[70] It is interesting that
the acceptor directionality is much more pronounced for
thiocarbonyl groups, with lone pair directions forming an
angle of only 105° with C=S (Figure 9, top).[69] The C=Se···H
angle in selenocarbonyl acceptors is even closer to rectangu-
lar.[8]

For hydroxy and water acceptors, the electronic structure
would predict a bimodal distribution with two preferred
directions in roughly tetrahedral geometry with respect to the

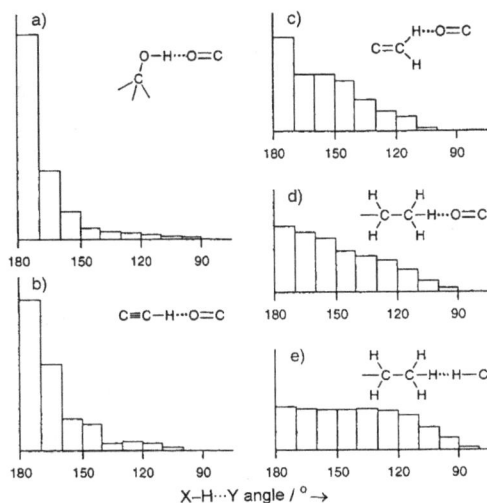

Figure 8. Directionality of X–H···O=C interactions with X–H groups of
different polarities (cone-corrected angular histograms). a) Hydroxy,
b) C≡C–H, c) C=CH₂, d) –CH₃ donors, e) C–H···H–C van der Waals
contacts. The degree of directionality decreases gradually from (a) to (d),
and interaction (e) is isotropic within a broad angular range. Note that the
picture for C≡C–H is similar to that for the conventional O–H donor.
C–H···O=C interactions of methyl groups are still directional, but to a
much smaller extent than C≡C–H···O=C hydrogen bonds.[22]

3.2. Acceptor Directionality

Hydrogen bonds are directional also at the acceptor side.
For strong hydrogen bonds (but only for these essentially), the
directionality of the acceptor corresponds to the geometry of
the covalent product obtained in a hypothetical proton-

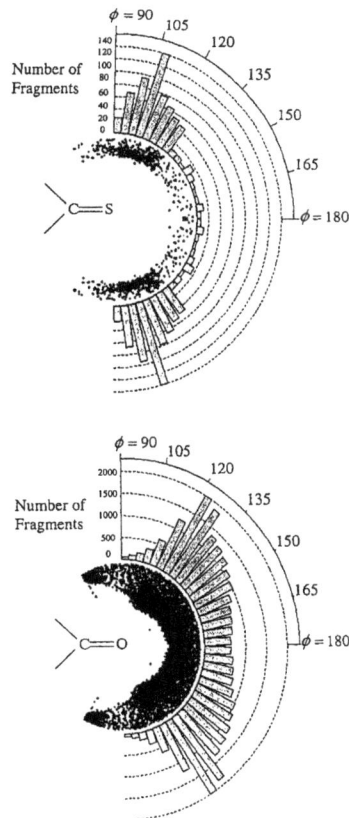

Figure 9. Acceptor directionality of C=O (bottom) and C=S (top) groups
in N/O–H···O/S=C hydrogen bonds.[69] Note that the directionality is
much more blurred for C=O than for C=S.

two covalent bonds at the O atom. The actual directionality, however, is so soft that clustering is only observed in the bisecting plane of R^1–O–R^2, without separation into two modes.[71] The acceptor directionality of the pyridyl N atom has also been characterized in a statistical study.[72]

Of more recent interest are hydrogen bonds with halogen acceptors. A metal-bonded halogen atom is strongly polar and a good acceptor. The electronic structure suggests different basicity characteristics of the different electron lone pairs (Scheme 3),[73] and indeed, X–H ··· Cl–M hydrogen bonds in crystals are almost exclusively donated roughly perpendicular to the M–Cl bond (angular range 80–140°).[73, 74] An exception among the metal-bonded halogens is fluorine, which shows a much more isotropic behavior.[75]

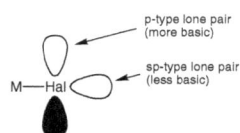

Scheme 3. The electron lone pairs of metal-bonded halogen.[73]

Most complex, and not yet fully explored, is the directionality of π acceptors. For the simplest one, C≡C, it seems that hydrogen bonds are preferably directed at the midpoint of the triple bond, but may point also at individual C atoms.[8] For the most important π acceptor, the Ph group, the potential energy surface of X–H ··· Ph interactions is very flat, which allows large movements of the donor over the π face without much of a difference in energy.[27i,k, 76] Consensus has not been reached concerning the location of the global energy minimum: does it occur with the X–H vector exactly over the ring midpoint (then, X–H can interact with *all* electrons of the π system), or does X–H point at an individual C–C bond or even a C atom? All these geometries, and also all intermediate situations, are found in crystals. Two extreme cases are illustrated in Figure 10 (for details, see legend).[77, 78] The large size of the π face makes the Ph acceptor a "target that is easy to hit".[79] This has important consequences for the role of the phenyl group in the packing of organic molecules,[8] and also for its role as a reserve acceptor in biological substances.[80] In the case of a local deficiency of conventional acceptors, a donor can form an X–H ··· Ph hydrogen bond instead if a Ph group is available even only in a roughly suitable geometry.

3.3. Distributions and Mean Values of the H ··· A Bond Lengths

Hydrogen bond lengths d in the solid state are very variable. Distances and angles vary in wide ranges even with a constant donor–acceptor combination, as shown already for the example C–OH ··· Cl$^-$ in the $d–\theta$ scatter plot shown in Figure 7a. Here, the broad scatter can not be a consequence of chemical variations, but only of crystal packing forces that affect each hydrogen bond in a different way (for a more detailed discussion of this matter, see Section 5).

Angew. Chem. Int. Ed. **2002**, *41*, 48–76

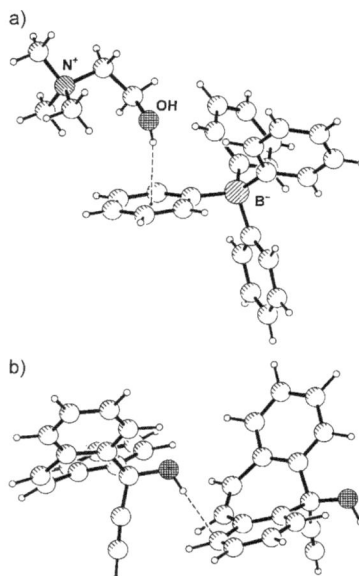

Figure 10. O–H ··· Ph hydrogen bonds with different geometries. a) The donor is positioned almost exactly over the aromatic midpoint M; the six H ··· C distances are in the range 2.49–2.70 Å; H ··· M is much shorter, 2.17 Å, and the O–H ··· M angle is 160° (X-ray crystal structure of choline tetraphenylborate).[77] b) The donor is oriented directly at a C atom, H ··· C = 2.34 Å, angle O–H ··· C = 174° (neutron diffraction crystal structure of 5-ethynyl-5*H*-dibenzo[*a*,*d*]cyclohepten-5-ol).[78]

If one wants to display bond lengths in a histogram, a cutoff in the angle θ has to be selected. If only linear hydrogen bonds are of interest, one may select $\theta > 135°$ and arrive at a distribution such as the one shown in Figure 11a for –N$^+$H$_3$ ··· Cl$^-$ hydrogen bonds. There is a distinct maximum, and the distribution has a well-defined beginning and a fairly

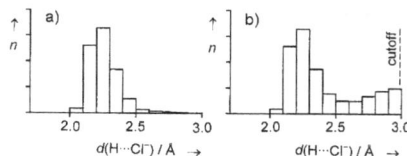

Figure 11. Typical H ··· A bond-length distributions: –NH$_3^+$ ··· Cl$^-$ hydrogen bonds up to a cutoff of $d = 3.0$ Å (data as for in Figure 7b). a) Only fairly linear hydrogen bonds with $\theta > 135°$. This histogram contains data from a horizontal slice $180 > \theta > 135°$ of Figure 7b. b) With the generous angle cutoff $\theta > 90°$. This histogram contains data from the slice $180 > \theta > 90°$ of Figure 7b.

well defined end with only a few outliers. For weak hydrogen bond types (or severely sterically hindered ones), the distribution does not fall to zero at long distances but fades into the continuum of random contacts.[81] If a more generous angle cutoff is chosen, such as $\theta > 110°$ or $90°$, the distance distribution will typically look like the one in Figure 11 b. Minor components of multifurcated interactions are now included, and as a consequence, the distribution no longer goes to zero at longer distances: following a certain minimum, the frequency of contacts increases again and merges with the continuum of random contacts. Since the relative content of multifurcated bonds is strongly sample-dependent (see Section 3.4), the exact shape of the long-distance region of such histograms is strongly sample-dependent too. Statistical

characterization of such distributions (such as Figure 11 b) is difficult.[82]

Figures 7 and 11 show the geometry variation for a constant donor – acceptor combination. If the donor and/or acceptor are chemically varied, new pictures are obtained which differ in the mean distance and the degree of directionality. Relatively comprehensive (and new) numerical data are compiled in Tables 3 and 4 for hydrogen bonds involving water molecules to shed light on the general rules determining mean hydrogen bond lengths.[10]

Table 3. Hydrogen bonds with water molecules as acceptors ($X–H \cdots O_W$): Geometry of fairly linear interactions ($\theta > 135°$) with various donors (distances are given in Å). (Database information was retrieved for this article.[10a])

Donor	n	Mean H $\cdots$ O_W distance	Mean X $\cdots$ O_W distance	H $\cdots$ O_W distance (95%)[a]	X $\cdots$ O_W distance (95%)[a]
O–H donors					
H_3O^+	21	1.54(2)	2.49(2)	–	–
N^+–OH	1	1.57(–)	2.55(–)	–	–
S–OH	4	1.58(–)	2.55(–)	–	–
P–OH	73	1.61(1)	2.575(9)	1.44 – 1.77	2.42 – 2.72
Se–OH	4	1.62(–)	2.59(–)	–	–
O=C–OH	244	1.629(4)	2.591(4)	1.51 – 1.78	2.49 – 2.75
N=C–OH	6	1.69(3)	2.60(3)	–	–
N–OH	46	1.68(1)	2.65(1)	–	–
C=C–OH, Ph–OH	162	1.724(8)	2.679(7)	1.55 – 1.96	2.52 – 2.88
As–OH	4	1.75(–)	2.68(–)	–	–
O–OH	2	1.76(–)	2.69(–)	–	–
C_{sp^3}–OH	763	1.804(4)	2.753(3)	1.64 – 2.06	2.61 – 2.97
$3(TM)OH^{[b]}$	6	1.81(5)	2.76(4)	–	–
$2(TM)OH^{[b]}$	14	1.85(4)	2.79(4)	–	–
H_2O	2505	1.880(2)	2.825(2)	1.72 – 2.19	2.68 – 3.11
B^-–OH	5	1.91(–)	2.86(–)	–	–
TM–OH[b]	5	1.96(–)	2.89(–)	–	–
$^-$OH	2	2.27(–)	3.22(–)	–	–
N–H donors					
$(SO_2,SO_2)NH$	7	1.71(1)	2.70(1)	–	–
Im^+N–H	20	1.74(2)	2.73(2)	–	–
Py^+N–H	67	1.78(1)	2.75(1)	1.63 – 2.05	2.63 – 2.96
$(C,C,C)N^+$–H	40	1.82(2)	2.77(1)	–	–
$(C,C)N^+H_2$	108	1.87(1)	2.83(1)	1.68 – 2.19	2.68 – 3.06
$(C_{sp^3},C_{sp^3})N$–H	316	1.860(8)	2.835(7)	1.69 – 2.20	2.69 – 3.13
C–NH_3^+	370	1.878(6)	2.840(5)	1.71 – 2.17	2.71 – 3.08
NH_4^+	86	1.95(1)	2.91(1)	1.74 – 2.24	2.73 – 3.11
$(C_{sp^2},C_{sp^2})N$–H	178	1.988(9)	2.937(8)	1.79 – 2.25	2.77 – 3.18
$(peptide)N$–H	118	1.99(1)	2.94(1)	1.80 – 2.31	2.77 – 3.18
C_{sp^2}–NH_2	508	2.016(6)	2.963(5)	1.81 – 2.31	2.78 – 3.21
$(TM,C,C)N$–H[b]	128	2.05(1)	2.99(1)	1.82 – 2.35	2.82 – 3.24
$(TM,C_{sp^3})N$–H[b]	18	2.07(3)	3.03(3)	–	–
$(TM,C)NH_2$[b]	467	2.084(6)	3.031(5)	1.88 – 2.35	2.86 – 3.27
TM–NH_3[b]	68	2.09(2)	3.03(1)	1.90 – 2.35	2.89 – 3.28
$(C_{sp^3},C_{sp^3})N$–H	13	2.14(3)	3.08(2)	–	–
C_{sp^3}–NH_2	20	2.12(4)	3.09(4)	–	–
N–NH_2	5	2.16(–)	3.09(–)	–	–
S–H donors					
C–SH	1	2.16(–)	3.48(–)		
C–H donors					
Cl_3C–H	2	2.06(–)	3.07(–)	–	–
$C≡C$–H	3	2.10(–)	3.16(–)	–	–
Cl_2CH_2	2	2.16(–)	3.22(–)	–	–
$(N,N)C_{sp^2}$–H	32	2.41(3)	3.38(3)	–	–
$(Cl,C)C_{sp^3}$–H	6	2.46(9)	3.44(5)	–	–
$(N,C)C_{sp^2}$–H	276	2.48(1)	3.47(1)	> 2.12	> 3.14
$(C,C)C_{sp^2}$–H	1369	2.553(4)	3.540(4)	> 2.22	> 3.23
$(C,C,C)C_{sp^3}$–H	29	2.59(4)	3.59(2)	–	–
O–CH_3	80	2.59(2)	3.59(2)	> 2.32	> 3.32
C_{sp^3}–CH_3	533	2.632(6)	3.613(6)	> 2.37	> 3.35

[a] The "95% ranges" of H $\cdots$ O_W and X $\cdots$ O_W distances include 95% of the hydrogen bonds. They are given only if $n > 50$. For distributions without a pronounced maximum, the 2.5th percentile is given instead of the central 95%. [b] TM = transition metal atom.

Solid-State Hydrogen Bonds   **REVIEWS**

Table 4. O_W–H $\cdots$ A hydrogen bonds from water donor molecules that are not coordinated to transition metal atoms. Geometry of fairly linear interactions ($\theta > 135°$) for various acceptors (distances in Å). (Database information was retrieved for this article.[10b])

Acceptor	n	Mean H $\cdots$ A distance	Mean $O_W \cdots$ A distance	H $\cdots$ A distance (95%)[a]	$O_W \cdots$ A distance (95%)[a]
O acceptors					
$^-$OH	8	1.71(3)	2.69(3)	–	–
Se=O	6	1.79(2)	2.74(2)	–	–
As=O	11	1.84(3)	2.76(2)	–	–
P=O, P–O$^-$	664	1.846(4)	2.793(4)	1.69 – 2.09	2.65 – 3.01
N$^+$–O$^-$	50	1.84(2)	2.80(1)	1.66 – 2.12	2.64 – 3.03
(C,C)C–O$^-$	95	1.85(1)	2.80(1)	1.62 – 2.17	2.59 – 3.11
–COO$^-$	1035	1.859(4)	2.807(3)	1.72 – 2.07	2.69 – 2.99
H$_2$O	2505	1.880(2)	2.825(2)	1.72 – 2.19	2.68 – 3.11
R$_2$C=O	2485	1.900(3)	2.840(2)	1.73 – 2.23	2.69 – 3.11
C$_{sp^3}$–OH	757	1.891(4)	2.831(4)	1.73 – 2.19	2.69 – 3.07
TM–O–C[b]	560	1.902(6)	2.842(6)	1.66 – 2.24	2.63 – 3.12
S=O, S–O$^-$	668	1.914(5)	2.854(4)	1.74 – 2.27	2.70 – 3.15
B–O–C	23	1.92(3)	2.86(2)	–	–
TM=O, TM–O$^-$ [b]	218	1.94(1)	2.877(8)	1.73 – 2.30	2.70 – 3.16
TM–O$_2$[b]	16	1.95(3)	2.88(2)	–	–
Ph–OH	89	1.97(1)	2.89(1)	1.72 – 2.27	2.62 – 3.17
C–O–C	254	1.978(9)	2.910(7)	1.78 – 2.33	2.74 – 3.17
N–OH	20	1.99(3)	2.91(2)	–	–
P–OH	34	1.97(2)	2.91(2)	–	–
NO$_3^-$	195	2.00(1)	2.927(9)	1.77 – 2.36	2.69 – 3.24
(O=C)–**OH**	35	2.01(3)	2.94(3)	–	–
Sb–O–C	20	2.03(4)	2.95(3)	–	–
ClO$_4^-$	180	2.07(1)	2.98(1)	1.80 – 2.36	2.73 – 3.25
Te–OH	5	2.07(–)	2.99(–)	–	–
C–NO$_2$	57	2.13(2)	3.04(2)	1.85 – 2.38	2.80 – 3.17
TM–CO[b]	4	2.30(–)	3.11(–)	–	–
N acceptors					
C$_{sp^3}$–NH$_2$	17	1.88(2)	2.84(1)	–	–
C$_{sp^3}$, C$_{sp^3}$NH	23	1.93(3)	2.89(2)	–	–
N=N–N	13	1.94(2)	2.89(2)	–	–
C$_{sp^3}$, C$_{sp^3}$, C$_{sp^3}$N	78	1.96(1)	2.90(1)	1.78 – 2.27	2.76 – 3.20
C=N–C	345	1.959(6)	2.905(6)	1.79 – 2.26	2.75 – 3.17
C=**N**–O	24	1.99(3)	2.94(2)	–	–
–C≡N	43	2.00(2)	2.94(2)	–	–
C$_{sp^3}$–NH$_2$	25	2.03(3)	2.95(2)	–	–
C ~ **N** ~ N	50	2.03(2)	2.96(2)	–	–
C=N–S	9	2.09(6)	3.03(5)	–	–
S acceptors					
C–S$^-$	68	2.38(1)	3.31(1)	2.22 – 2.61	3.19 – 3.51
P=S, P–S$^-$	12	2.35(2)	3.31(1)	–	–
Sn–S$^-$	7	2.41(2)	3.33(3)	–	–
R$_2$C = S	73	2.42(1)	3.36(1)	2.26 – 2.65	3.24 – 3.58
TM–S–C[b]	16	2.51(3)	3.43(3)	–	–
C–S–C	2	2.60(–)	3.53(–)	–	–
Se acceptors					
Se	3	2.45(–)	3.40(–)	–	–
Halogen acceptors					
F$^-$	13	1.70(2)	2.67(2)	–	–
SiF$_6^{2-}$	12	1.84(2)	2.79(2)	–	–
TM–F[b]	45	1.85(3)	2.80(2)	–	–
BF$_4^-$	34	2.01(3)	2.94(3)	–	–
PF$_6^-$	18	2.08(3)	2.98(3)	–	–
C–F	5	2.19(–)	3.07(–)	–	–
Cl$^-$	1013	2.245(3)	3.196(3)	2.10 – 2.46	3.06 – 3.38
TM–Cl[b]	232	2.349(9)	3.272(8)	2.15 – 2.62	3.11 – 3.51
C–Cl	30	2.77(5)	3.62(5)	–	–
Br$^-$	233	2.415(8)	3.350(7)	2.25 – 2.66	3.21 – 3.61
TM–Br[b]	17	2.56(4)	3.47(4)	–	–
C–Br	1	2.83(–)	3.66(–)	–	–
I$^-$	47	2.68(1)	3.61(1)	–	–
TM–I	6	2.90(8)	3.74(6)	–	–
π acceptors					
Ph	25	2.50(4)	3.38(4)	–	–
C≡C	2	2.51(–)	3.35(–)	–	–
C=C	20	2.73(4)	3.57(4)	–	–
Py	4	2.79(–)	3.72(–)	–	–

[a] The "95% ranges" of the H $\cdots$ A and $O_W \cdots$ A distances include 95% of the hydrogen bonds. They are given only if $n > 50$. For distributions without a pronounced maximum, the 2.5th percentile is given instead of the central 95%. [b] TM = transition metal atom.

REVIEWS

Mean distances are listed in Table 3 for fairly linear X–H $\cdots$ O$_W$ hydrogen bonds (W = water molecule) from 47 X–H donor types (X = O, N, S, and C). If one uses the ranking of distances to define a "donor strength" of X–H, one finds a general ranking O–H > N–H > S–H > C–H. These are only rough categories, however, with strong internal variations. Hydrogen bonds of the strongest C–H types (Cl$_3$CH, C≡CH) are clearly shorter on average than those with the weakest N–H donors (C$_{sp^2}$–NH$_2$, N–NH$_2$). The ranking within the X–H groups follows a simple rule: basically, the donor strength is increased by neighboring electron-withdrawing groups and reduced by electron-donating groups. In consequence, the ranking of O–H donor strengths is H$_3$O$^+$ > O=C–OH > Ph–OH > C$_{sp^3}$–OH > H$_2$O > OH$^-$. The difference in mean bond lengths D within this sequence amounts to over 0.7 Å! Remember, however, that each of the lines in Table 3 corresponds to a broad histogram (such as in Figure 11 a). The corresponding 95 % ranges are normally over 0.3 Å broad (last two columns in Table 3), which implies there are large overlapping regions of hydrogen bond geometries between most donor types.

An analogous list is given in Table 4 for water O$_W$–H $\cdots$ A hydrogen bonds with 61 different acceptor types (A = O, N, S, Se, halogen, π system). It is clear that acceptor strengths are increased by neighboring electron-donating groups and reduced by electron-withdrawing groups. The ranking of strengths for O acceptors is $^-$OH > $^-$COO$^-$ > H$_2$O > C$_{sp^3}$–OH > Ph–OH > C–NO$_2$ > M–CO, with mean bond lengths varying by 0.6 Å. A broad range of acceptor strengths is also observed for fluorine: F$^-$ > M–F > BF$_4^-$ > C–F.

Tables 3 and 4 summarize the data for H$_2$O as an acceptor and a donor, respectively. A more comprehensive picture would require data for *all* donor–acceptor combinations. The X–H and A groups from Tables 3 and 4 would fill a 48 × 61 matrix with almost 3000 entries, too many to be discussed in practice. Sections of this matrix have been published for the special cases of X–H $\cdots$ Hal$^-$,[60] C–H $\cdots$ O and C–H $\cdots$ N,[83] and C–H $\cdots$ π[84] hydrogen bonds.

General properties of donor–acceptor matrices are best discussed with smaller example matrices, such as those in Tables 5 and 6. Part of the O–H $\cdots$ O matrix with four donors, and four acceptors that cannot act as donors simultaneously (C=O, etc.) is shown in Table 5. The ranking of the donor strength is here independent of the acceptor, and the ranking of the acceptor strength is independent of the donor.

A related matrix is shown in Table 6 for four kinds of O–H groups acting as donor as well as acceptor. Here, the important observation is that strong donors are weak acceptors, and vice versa. The O–H group of the carboxylic acid, for

Table 6. Intermolecular O–H $\cdots$ O hydrogen bonds: Donor–acceptor matrix with O–H groups that can act as donors as well as acceptors (mean O $\cdots$ O distances in Å, sample size in square brackets). (For this article database information was retrieved.[10a])

Donor	Acceptor			
	H–O–H	C$_{sp^3}$–O–H	Ph–O–H	(O=C)–**O**–H
H–O–H	2.825(2) [2505]	2.831(4) [757]	2.89(1) [89]	2.94(3) [35]
C$_{sp^3}$–O–H	2.753(3) [763]	2.792(2) [4249]	2.84(1) [94]	2.89(–) [3]
Ph–O–H	2.679(7) [144]	2.721(7) [145]	2.807(6) [305]	2.97(–) [2]
O=C–O–H	2.591(4) [244]	2.646(6) [162]	2.69(2) [8]	–

example, is a very strong donor but a very poor acceptor. The water molecule is a good acceptor but only a moderate donor.

The distance properties d and D discussed above might indicate that donor and acceptor "strengths" are integral properties of any group X–H or A. The shortest hydrogen bonds could then be made by simply combining the strongest donor with the strongest acceptor. Such a view is correct in the electrostatic regime of hydrogen bonds, namely, for those defined as "moderate" and "weak" in Section 2.10. It is *not* correct, however, for strong hydrogen bonds, where the laws of covalent bonding and of proton transfer phenomena become dominant (see Section 7). For example, if one tries to make a short O–H $\cdots$ O bond by simply combining the strongest donor from Table 3 (H$_3$O$^+$) with the strongest acceptor in Table 4 (OH$^-$) a proton transfer will occur (H$_2$O$^+$–H $\cdots$ OH$^-$ →H$_2$O $\cdots$ H–OH) which leads to a moderate hydrogen bond between water molecules. Similarly, it would be wrong to assume a strict relationship between hydrogen bond length and energy. Hydrogen bonds involving ions generally have a higher dissociation energy than those between neutral molecules (this is a trivial consequence of the Coulombic attraction of the net charges), but need not have much shorter bond lengths (see, for example, the charged acceptor NO$_3^-$ in Table 4, with a long average O $\cdots$ O$_W$ distance of 2.97 Å).

3.4. Bifurcated Hydrogen Bonds

In a multifurcated hydrogen bond, a donor forms hydrogen bonds with more than one acceptor simultaneously (Scheme 2). Multifurcated hydrogen bonding requires a high density of acceptors, at least locally (Scheme 2). Over 25 % of all O–H $\cdots$ O hydrogen bonds in carbohydrates are multifurcated, and this fraction is even higher in amino acids.[6] Proteins also contain multifurcated hydrogen bonds in large numbers.[85] It is not always easy to show that all components are bonding, in particular if angles θ are small and/or some of the putative acceptors are forced by stereochemistry to be close to the donor. For a number of bifurcated bonds, however, bond paths in the theoretical electron density have been shown for both components.[86]

As a consequence of their geometry, certain chemical groups are involved in these interactions with a particularly high frequency. A typical example of the *ortho*-dimethoxyphenyl group is shown in Scheme 4. In a database analysis,[10c] 31 O–H $\cdots$ O hydrogen bonds with this group were found, and of these, only 10 involve just one acceptor, whereas 21

Table 5. Intermolecular O–H $\cdots$ O hydrogen bonds with acceptors that cannot act as donors (mean O $\cdots$ O distances in Å, sample size in square brackets). (Database information was retrieved for this article.[10a])

Donor	Acceptor			
	–COO$^-$	R$_2$C=O	C–O–C	C–NO$_2$
O=C–OH	2.544(3) [421]	2.644(1) [1491]	2.72(2) [29]	2.80(–) [3]
Ph–OH	2.65(1) [57]	2.734(5) [412]	2.812(2) [58]	2.96(3) [11]
C$_{sp^3}$–OH	2.736(5) [354]	2.824(2) [2567]	2.885(4) [764]	3.00(1) [74]
H–O–H	2.807(3) [1035]	2.840(2) [2485]	2.910(7) [254]	3.04(2) [57]

62

Angew. Chem. Int. Ed. **2002**, *41*, 48–76

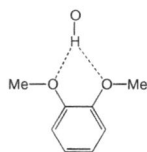

Scheme 4. Example of a functional group with a strong tendency to accept bifurcated hydrogen bonds (21 out of 31 hydrogen bonds found in a CSD analysis were bifurcated).[10c]

(=68%) are bifurcated. Of the latter, 8 are almost symmetric, with the two H ··· O distances differing by less than 0.2 Å.

A strongest ("major") component can be clearly identified in most multifurcated hydrogen bonds, but not always. Even trifurcated hydrogen bonds occur occasionally with a fairly symmetric geometry. The triethanolammonium cation, for example, is always found in crystal structures in very similar bowl-shaped conformations with the three hydroxy O atoms converging toward the N⁺−H donor (Figure 12).[77, 87]

Frequently, the two acceptors of a bifurcated hydrogen bond are of different types, A_1 and A_2. If one is much weaker than the other (such as with $A_1 = O$, N, and $A_2 = Hal-C$, π, etc.) it may be difficult to assess if it is actually of any structural importance. However, numerous examples of bifurcated hydrogen bonds have been found with a strong and a weak acceptor, occasionally even with the interaction geometry more favorable to the weaker acceptor (Scheme 5).[88]

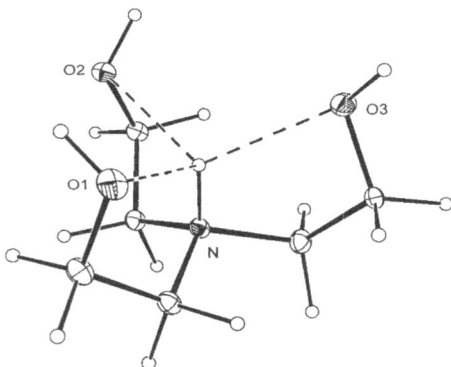

Figure 12. Trifurcated hydrogen bond in the triethanolammonium cation as seen in the neutron diffraction crystal structure of the dihydrated tetraphenylborate salt.[87] The three N⁺−H ··· O hydrogen bonds have very similar geometries (H ··· O = 2.14–2.35, N ··· O = 2.71–2.86 Å, N−H ··· O = 108–112°).

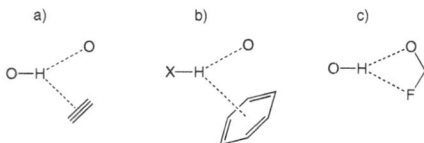

Scheme 5. Examples of bifurcated hydrogen bonds with a strong and a weak acceptor: a) with O and C≡C acceptors;[8, 88a,b] b) with O and Ph acceptors;[80, 88c] c) with O and F−C acceptors.[88d]

3.5. H···H Contacts

The matter of short repulsive H ··· H contacts is often overlooked when interpreting hydrogen bond geometries. If a hydrogen bond is formed between two X−H groups (or an X−H and an Y−H group), the two may be roughly in-line so that the H atoms are far apart from each other, but they may also form an angle in such a way that the H atoms approach quite closely (Figure 13). In database analyses of inorganic[89] and organic[90] crystal structures, a lower limit of 2.05 Å was

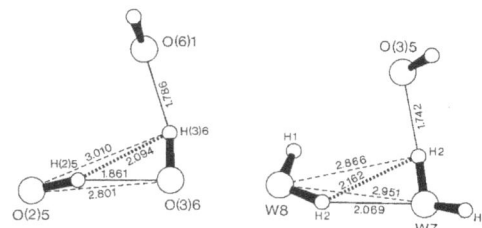

Figure 13. Typical examples of short H ··· H contacts in O−H ··· O−H hydrogen bonds found by neutron diffraction studies of carbohydrates.[90]

found for H ··· H contacts in such configurations. This does not affect linear hydrogen bonded chains very much, but it imposes serious constraints on the geometry of circular arrays of hydrogen bonds. Short H ··· H contacts cannot be avoided in rings of three hydroxy groups or water molecules in particular (Scheme 6). They force angles θ to be very bent, and probably are the reason why these rings are very rare. Fairly short H ··· H contacts occur also in cyclic hydrogen-bonded dimers and destabilize such arrays, for example, in the carboxylic acid dimer with H ··· H contacts of about 2.34 Å (Scheme 7) and many related patterns.

Scheme 6. Short H ··· H contacts are unavoidable in rings of three O−H ··· O hydrogen bonds. The mean geometry in crystals is: $d = 2.07(3)$ Å, $D = 2.89(2)$ Å, $\theta = 143(2)°$, H ··· H = 2.21(3) Å.[10d]

3.6. Influence on Covalent Geometry

Hydrogen bonding affects the covalent geometry of the molecules involved. The lengthening of the covalent X−H bond was already described in the 1950s,[91] and the correlation of the O−H and H ··· O distances in O−H ··· O interactions has been studied many times with increasing precision. The current correlation based on low-temperature neutron diffraction data is shown in Figure 14a.[92] The O−H bond continuously elongates with decreasing H ··· O distance until a symmetric geometry O−H−O is reached at an O ··· O separation of about 2.39 Å.

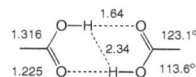

Scheme 7. Mean geometry of the carboxylic acid dimer in crystals.[10e] Note the short destabilizing H ··· H contact. The mean O ··· O distance is 2.644 Å.

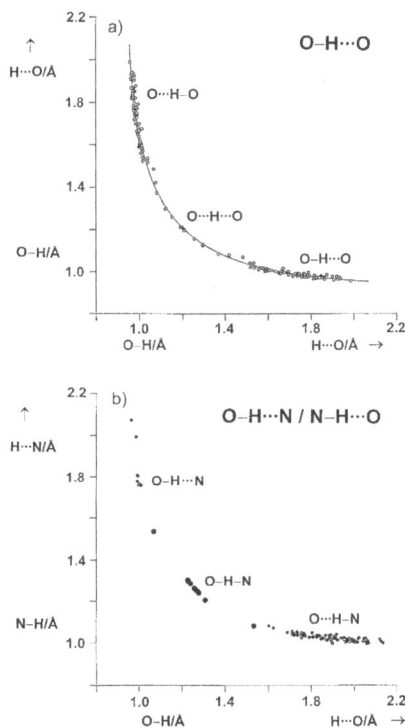

Figure 14. Lengthening of the X–H bond in X–H ··· A hydrogen bonds. a) Correlation of O–H and H ··· O distances in O–H ··· O hydrogen bonds.[92] The plot is symmetrized with respect to the two O atoms.[31b] b) Correlation of N–H with H ··· O bond lengths and O–H with H ··· N bond lengths.[93] The right branch shows N–H ··· O, and the left branch O–H ··· N hydrogen bonds. Both plots are based on neutron diffraction data.

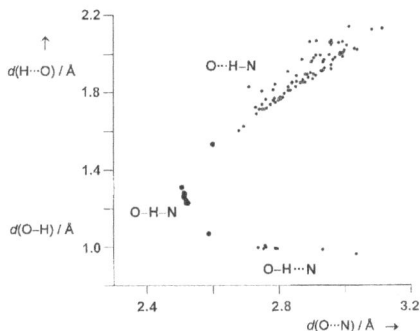

Figure 15. Correlation of O–H and H ··· O bond lengths with the O ··· N distance in O/N–H ··· N/O hydrogen bonds (neutron diffraction data).[93]

The correlation is perfectly smooth. There is no indication of a critical distance at which the hydrogen bond switches from one interaction type to another. The elongation is in the range 0.02 – 0.08 Å for moderate hydrogen bonds (Table 2), but it is up to 0.25 Å for strong ones. The analogous scatter plot for N–H ··· O and O–H ··· N hydrogen bonds also shows a smooth correlation (Figure 14b),[93] with the geometrically symmetric bond occuring at an N ··· O distance of about 2.50 Å. An alternative way to illustrate the elongation of X–H bonds is to draw X–H and H ··· A distances as a function of the X ··· A separation, as shown in Figure 15 for the example of O/N–H ··· N/O bonds.

Lengthening of the X–H bond has been found for many other kinds of X–H ··· A hydrogen bonds, and seems typical of hydrogen bonding. The effect has been described for N–H ··· N,[94] O–H ··· S,[36] N–H ··· S,[36] O–H ··· Cl⁻,[36] N–H ··· Cl⁻,[36] and C–H ··· O[95] interactions, although the full range of geometries has been explored only for the types shown in Figure 14. For some special systems,

quantum chamical calculations predict shortening, not lengthening, of the X–H bond,[55, 96] but there is no consensus on this among theorists[271] and experimental structural proof is still lacking.

X–H ··· A hydrogen bonding also affects the angles at X (Scheme 8). For the example of –NH₃⁺ ··· Cl⁻ hydrogen bonds, the bending of the C–N–H angle as a function of the

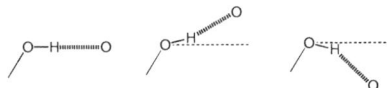

Scheme 8. Bending of the X–O–H angle in O–H ··· A hydrogen bonds.

C–N⁺ ··· Cl⁻ coordination angle is shown in Figure 16.[36] The change in the angle between the covalent bonds follows that in the coordination angle, but is much smaller. Related plots have been given for the bending of C–O–H groups, and for the opening and narrowing of the water angle by O–H ··· O and O–H ··· Cl⁻ hydrogen bonds.[97] The bending typically amounts to only a few degrees, which is easy to see with neutron diffraction, but difficult to discern in X-ray diffraction studies.

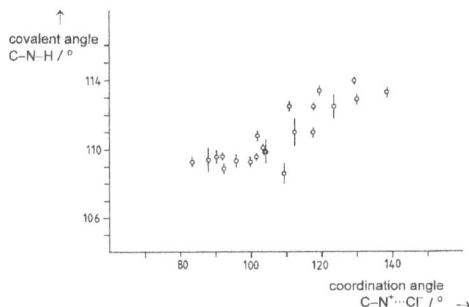

Figure 16. Bending of the C–N–H angle in N–H ··· Cl⁻ hydrogen bonds (–NH₃⁺ donors, neutron diffraction data).[36]

Snow, Ice and Other Wonders of Water

REVIEWS

Torsional angles in molecular fragments may be influenced by hydrogen bonds to a degree that depends on the height of the torsional barrier between energy minima. The O–H bond in a hydroxy group C_{sp^3}–OH, for example, prefers to be staggered with repect to the substituents at C (Scheme 9a). In

a)   b)

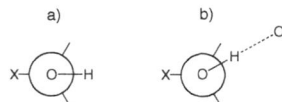

Scheme 9. Effect of hydrogen bonding on torsional angles. a) Free, and b) hydrogen-bonding hydroxy group.

crystal structures, on the other hand, the torsion angle around C–O takes any value. If this angle is plotted against the X–C–O ··· O dihedral coordination angle, it is seen that the O–H vector is always rotated towards the acceptor, even into an eclipsed orientation (Scheme 9b and Figure 17).[98] The

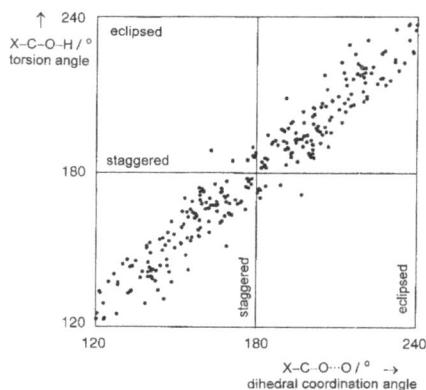

Figure 17. Rotation of hydroxy groups away from a staggered conformation as a result of hydrogen bonding: the X–C–O–H torsion angle adjusts to the X–C–C ··· O dihedral coordination angle.[98]

same kind of analysis can be performed for methyl groups of the type C_{sp^3}–CH$_3$ involved in C–H ··· O interactions. In this case, the torsional barrier is far too high (typically 3–5 kcal mol^{-1}) to reach the eclipsed conformation. Nevertheless, rotations of up to 10–15° from the ideally staggered conformations have been detected, which corresponds to a displacement of the H atoms by 0.2 Å from their idealized positions.[98] A single case of an eclipsed methyl group associated with three C–H ··· O bonds has been reported,[99] but theoretical calculations show that the torsional barrier in this particular molecule is dramatically reduced to about 1.5 kcal mol^{-1}.[100]

The effects discussed above concern distances and angles directly involving the H atom. There are also changes in the covalent geometry of the non-hydrogen molecular skeleton of

both participating molecules. These effects are modest for moderate, and negligible for weak hydrogen bonds, but they can become very large for strong ones. In a C–O–H ··· O=C hydrogen bond, for example, the C–O bond is shortened and the O=C bond is lengthened compared to the free molecules.[101] In the extreme case of the symmetric hydrogen bond, the two C–O bond lengths become identical, C–O ··· H ··· O–C. Figure 18 shows that the difference of the two C∼O distances depends linearly on the O ··· O distance (note that the effect is already quite large with O ··· O = 2.6 Å).[46] Carboxylate groups that accept one hydrogen bond become unsymmetric, and the C∼O bond involving the accepting O atom becomes several hundreths of an Å longer than the other one (see, for example, ref. [102]).

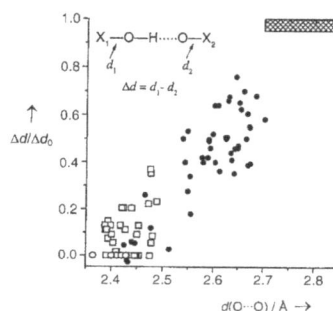

Figure 18. Effect of C–O–H ··· O=C hydrogen bonding on C∼O bond lengths. The quantity Δd is the difference between the donor C–O and the acceptor O=C bond length (d_0 = value for fragments free of hydrogen bonding; squares: combination of acid and complementary base; dots: resonance-assisted hydrogen bonds).[46]

All these correlations can be easily rationalized if they are interpreted as mapping a proton-transfer reaction. The geometry of the donor molecule changes in the direction of the deprotonated species, and the geometry of the accepting molecule changes in the direction of a protonated one. This has been nicely demonstrated by suitable chemical variations of substituted phenol–amine adducts[103] to form molecular or ionic adducts, and also into intermediate cases.[104] A plot of the phenolic C–O bond length against the O ··· N distance gives the correlation shown in Figure 19. The data in the upper right corner show normal phenolic C–O bond lengths of around 1.34 Å, and are associated with moderate strength O–H ··· N hydrogen bonds. As the hydrogen bonds become shorter, the H atom is gradually abstracted and the C–O bond shortens. The data in the bottom right corner show a phenolate C–O bond length of 1.25 Å, and are associated with ionic hydrogen bonds N$^+$–H ··· O$^-$–C. The symmetric situation C–O ··· H ··· N is reached at an N ··· O distance of about 2.50 Å.

A related effect is observed for the C∼N∼C angle in pyridine molecules (Figure 20).[10f] The neutral pyridine molecule has an angle at the nitrogen atoms of 116.6°,[105] whereas in pyridinium ions they are widened to 121–123° (for example, 122.6° in Py · HCl · HCl[106]). This angle in O–H ··· N

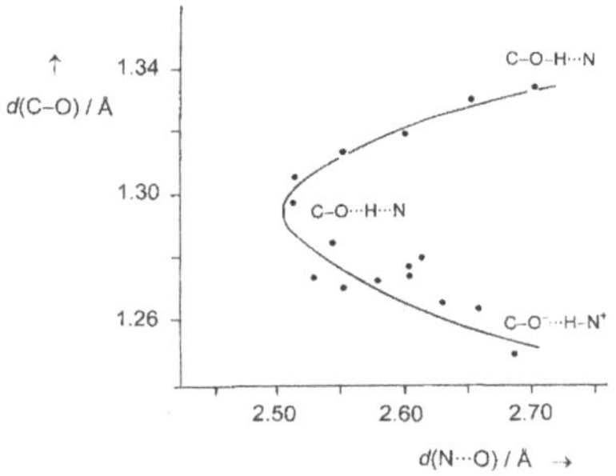

Figure 19. Effect of hydrogen bonding on the phenolic C–O bond length. Data from adducts of phenols and amines.[103]

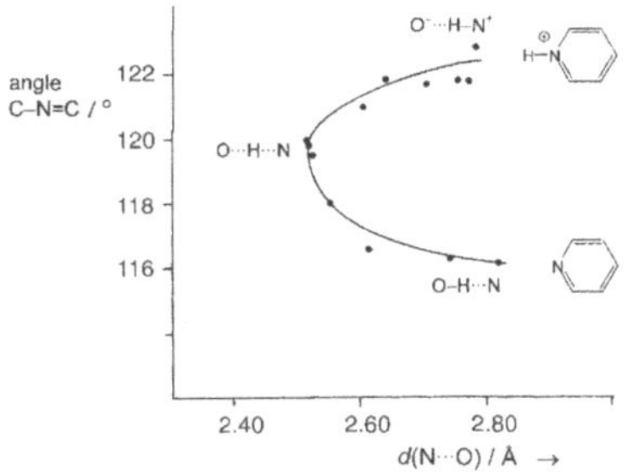

Figure 20. Effect of hydrogen bonding on the C–N=C angle in pyridine.[10f]

hydrogen bonds is increasingly widened as the hydrogen bond becomes shorter (lower branch of the curve), and it is narrowed in N^+–H ··· O hydrogen bonds as the bond becomes shorter. The branches meet again at O ··· N = 2.50 Å, with an angle at N of 120°. Changes in covalent geometry also play an important role in the case of resonance-assisted hydrogen bonding (see Section 4.2).

3.7. H/D Isotope Effects

The H/D isotope effect is a curious matter in the area of hydrogen bonds. In the classical Ubbelohde effect, hydrogen bond lengths slightly increase upon deuteration.[107] This is thought to be a result of the lower zero-point vibrational energy of the O–D relative to the O–H bond, which makes the O–D bond more stable. In consequence, D is more difficult to abstract from O than H, and the hydrogen bonds are weaker. The Ubbelohde effect has been examined experimentally for only a few O–H/D ··· O pairs. From a recent survey it appears that the isotope effect is strongest in the O ··· O distance range 2.5–2.6 Å (Figure 21).[108] It is much smaller for long hydrogen bonds, and about zero for very short ones (possibly, even negative values are allowed). The details of the effect are as yet unexplained.

Isotope exchange occasionally leads to more severe structural changes. A well-studied example is trifluoroacetic acid

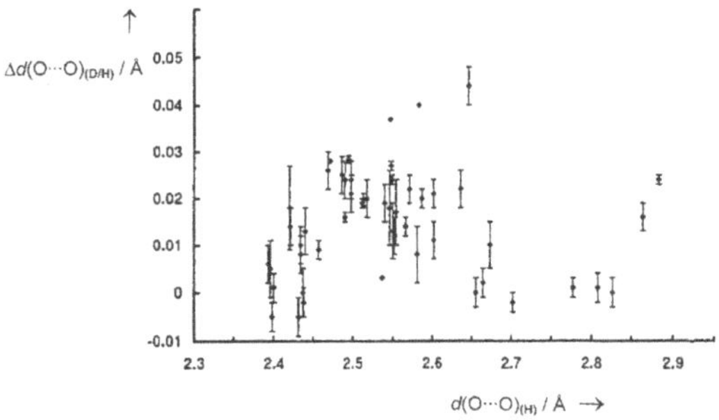

Figure 21. H/D isotope effect on hydrogen bond lengths. The difference of O ··· O distances in pairs of D and H compounds is plotted against the O ··· O distance of the H compound. A value of $\Delta > 0$ means that the hydrogen bond in the D compound is longer.[108]

tetrahydrate, which is cationic in the protonated form, F_3C–$COO^- \cdot 3H_2O \cdot H_3O^+$, whereas it becomes molecular when deuterated, F_3C–$COOD \cdot 4D_2O$.[109]

4. Non-Additivity

Many properties of *n* interconnected hydrogen bonds are not just the sum of those of *n* isolated bonds. Two principal mechanisms are responsible for this non-additivity, and both operate by mutual polarization of the involved groups.

4.1. σ-Bond Cooperativity

If an $X^{\delta-}$–$H^{\delta+}$ group forms a hydrogen bond $X^{\delta-}$–$H^{\delta+} \cdots A^{\delta-}$, it becomes more polar. The same is true if it accepts a hydrogen bond, $Y^{\delta-}$–$H^{\delta+} \cdots X^{\delta-}$–$H^{\delta+}$. Thus, in a chain with two hydrogen bonds, Y–H ··· X–H ··· A, *both* become stronger. The effect is often called "σ-bond cooperativity",[6] since the charges flow through the X–H σ bonds, but the terms "polarization-enhanced hydrogen bonding"[110] or "polarization-assisted hydrogen bonding" (as opposed to resonance-assisted hydrogen bonding)[46] have also been proposed. Model calculations on moderate strength hydrogen bonds yield typical energy gains of around 20% relative to isolated interactions.[7]

σ-Bond cooperativity drives the clustering of polar groups. In the condensed phases, this leads to formation of X–H ··· X–H ··· X–H chains and rings, in particular for X = O, but also for X = N or S. If double donors (such as H_2O) and/or double acceptors are involved, they can interconnect chains and rings to form complex networks. The topology of such networks has been documented in great detail for the O–H-rich carbohydrates.[5, 111, 112]

An unusual array cooperative σ-bond is shown in Scheme 10. The polarity of the water molecule is greatly enhanced by two hydrogen bonds donated to strong O=P

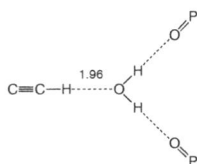

Scheme 10. A very short C≡C–H ··· O hydrogen bond. σ-Bond coorperativity enhances the acceptor strength of the water molecule (C ··· O = 3.02 Å).[113]

acceptors, so that the accepted C≡C–H ··· O hydrogen bond becomes very short, in fact the shortest ever found for an acetylenic donor.[113]

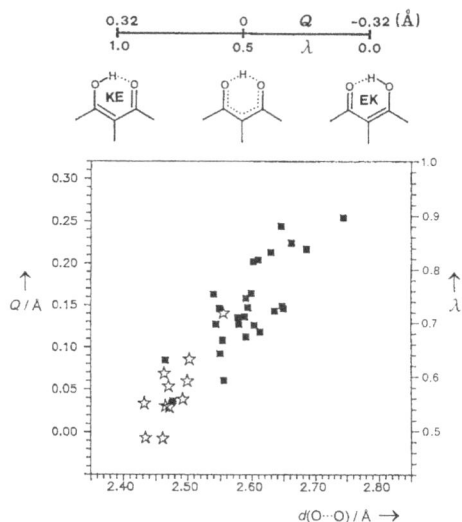

4.2. π-Bond Cooperativity or Resonance-Assisted Hydrogen Bonding (RAHB)

X–H groups may also be polarized by charge flow through π bonds. For example, an amide N–H group becomes a stronger donor if the amide O atom accepts a hydrogen bond, X–H ··· O=C–N–H. This results because the zwitterionic resonance form is stabilized (Scheme 11). The same effect occurs in thio- and seleno-amides.[8] Amide units, as a result of their dual donor and acceptor capacity, often form hydrogen-bonded chains or rings (such as in protein secondary structure; Scheme 12). Since the polarization occurs through π bonds, the effect is often called π-bond cooperativity.[6]

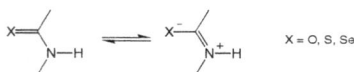

Scheme 11. Resonance forms of amide, thioamide, and selenoamide groups. The neutral form is always dominating, but the weight of the zwitterionic form is increased by accepted as well as by donated hydrogen bonds.

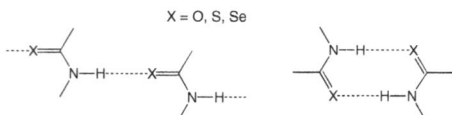

Scheme 12. Chains and rings as formed by amides, thioamides, and selenoamides through the π-bond cooperativity.

On the basis of studies of intramolecular hydrogen bonds in β-diketone enolates Gilli et al. call this effect "resonance-assisted hydrogen bonding" (RAHB).[114] A short hydrogen bond in the β-diketone enolates is associated with a charge flow through the system of conjugated double bonds (Scheme 13). The C–O and C–C bonds gain partial double bond character and are shortened, whereas the C=O and C=C

Scheme 13. Resonance-assisted hydrogen bonding (RAHB) in β-diketones enolates.[114]

bonds are weakened correspondingly. If a delocalization parameter $Q = (d_1 - d_4) + (d_3 - d_2)$ is plotted against the O ··· O distance (Figure 22) it is seen that the delocalization systematically increases with a shortening of the hydrogen bond length. In the extreme case of a symmetric position of the H atom, Q is zero and the entire fragment becomes symmetric. Completely analogous effects operate in

Figure 22. Resonance-assisted hydrogen bonding (RAHB) in enolones according to Gilli et al.[114] The parameter Q measuring the degree of π delocalization decreases with decreasing O ··· O distance ($Q = 0$ indicates a completely delocalized π system). Stars represent intramolecular, and squares represent intermolecular hydrogen bonds.

longer chains of conjugated double bonds with intra- and also intermolecular hydrogen bonds.[115] The best known example is the carboxylic acid dimer (Scheme 7). Any other suitable donor–acceptor pair connected by a resonant π system will also show the effect. The cases N–H ··· O and O–H ··· N,[116] N–H ··· S/Se,[8] O–H ··· S,[117] and S–H ··· S[118] (Schemes 12 and 14) illustrate the variety, for which experimental evidence of the π-bond cooperativity is available.

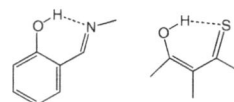

Scheme 14. Examples of π-bond cooperative (or resonance-assisted) hydrogen bonds other than O–H ··· O. Further examples are given in Scheme 12.

4.3. Anticooperativity

Hydrogen bonds may not only enhance, but also reduce the strengths of each other. This occurs, for instance, at double acceptors where two roughly parallel donor dipoles repel each other (Scheme 15). This effect is probably responsible for the preferences of "homodromic" over "antidromic" cycles of

Angew. Chem. Int. Ed. **2002**, *41*, 48 – 76

67

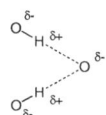

Scheme 15. Anticooperative hydrogen bonds. The two donors represent roughly parallel dipoles that repel each other.

hydrogen bonds (Scheme 16).[119] Given their importance for determining intermolecular structures, anticooperative effects have been investigated surprisingly little until now.

5. Elongation and Compression by Other Forces

Hydrogen bonds rarely adopt optimal geometry in the condensed states. Crystal-packing forces can easily bend, elongate, and compress them. All these distortions are associated with a weakening of the interaction (typically by only a few tenths of a kcal mol^{-1}, and only rarely above 1 kcal mol^{-1}). It is a common mistake to believe that in a set of hydrogen bonds between chemically identical groups, the shortest are the strongest. This is *not* the case. In such a set, the strongest hydrogen bonds are those closest to optimal geometry, whereas the shortest may well suffer some compression.

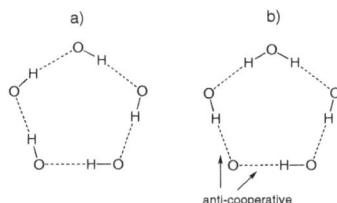

Scheme 16. Cycles of five hydrogen bonds.[119] In the preferred "homodromic" arrangement (a) all hydrogen bonds run in the same direction. In the less common "antidromic" arrangement (b) a change of orientation leads to local anticooperativity.

For moderate and strong hydrogen bonds, the hydrogen bond potential is deep enough (< -4 kcal mol^{-1}) to keep the distortions relatively limited (Figure 23). The distribution of geometries is narrow enough to show recognizeable clustering (see, for example, Figures 7 and 11). Even the most compressed contacts are deep in the bonding regime. For weak hydrogen bonds with dissociation energies of ≤ 1 kcal mol^{-1}, the entire range of bonding geometries is accessible, and compressed contacts may even be pushed up into the destabilizing regime (Figure 23c).[8]

For moderate and strong hydrogen bonds, statistical distance distributions often resemble a Morse potential drawn upside down (compare Figures 1 and 11). It is tempting to interpret such a distance distribution as Boltzmann population of a hydrogen bond potential. This would mean that the distortions of a particular flexible moiety (in this case a hydrogen bond) in the crystal fields of a large number of crystal structures statistically behave the same way as the distortions experienced in the thermal bath of a solution. This view has actually been taken by a number of authors.[71, 72, 120] However, quantitative calculations require introduction of quite arbitrary additional assumptions to obtain a reasonable

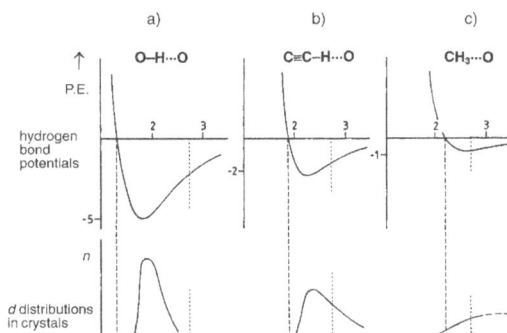

Figure 23. Schematic potential energy curves and distance distributions in crystals for three types of hydrogen bonds. Most hydrogen bond energies are within 1 kcal mol^{-1} of the minimum. The destabilizing region is accessible only for very weak hydrogen bond types. The H ··· O van der Waals separation of 2.7 Å is also indicated, and illustrates that the van der Waals cutoff definition is quite harmless for stronger hydrogen bond types, whereas it omits a large part of the weak hydrogen bonds.[8]

fit of model populations with experimental distance distributions.[120e] Distance distributions do not resemble Boltzmann population of a Morse potential at all for weak hydrogen bonds (Figure 23).[8, 81] Strong criticism was raised by Bürgi and Dunitz, who state that the approach is fundamentally wrong because "an ensemble of structural parameters obtained from chemically different compounds in different crystal structures does not even remotely resemble a closed system at thermal equilibrium and does not therefore conform to the conditions necessary for the application of the Botzmann distribution".[121]

In Sections 3 and 4, several effects have been discussed that cause a shortening of hydrogen bonds associated with real strengthening (increasing the polarity of X–H or A, cooperativity, etc.). Such effects can be recognized only from large samples of structural data in which effects from random scatter are smaller than the systematic trends. In individual crystal structures, "random distortions" are always chemically rooted although they may be difficult to explain.

6. Probability of Formation

It is difficult to predict whether or not a hydrogen bond between a potential donor X–H and a potential acceptor A in a given system will actually be formed. One can only define a probability of formation, that is, the fraction of X–H ··· A hydrogen bonds among the number of such hydrogen bonds that could be formed in principle. This is a global property of the sample, averaged over all chemical and structural situations. Nevertheless, the probability of formation is important for judging if a given type of hydrogen bond is general or exotic. Only if a hydrogen bond (or an array of hydrogen bonds) has a reasonably high probability of formation can it be used in rational crystal design ("crystal engineering"). Hydrogen bonds that are formed only occasionally, will be very difficult to control (although they must be understood for rationalizing a crystal structure they are found in).

Probabilities of formation can be determined for single hydrogen bonds, as well as for hydrogen bond arrays. In a pioneering investigation work, Allen and co-workers have determined the global probabilities of formation of 75 bimolecular ring motifs;[122] part of the results is shown in Scheme 17. The surprisingly poor performance of certain

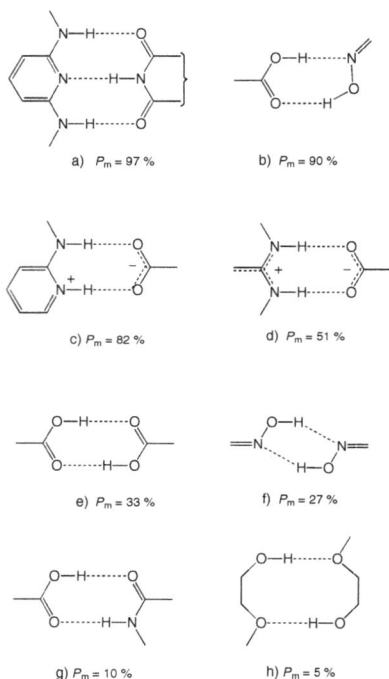

Scheme 18. Three kinds of intramolecular O–H ··· O hydrogen bonds with their probability of formation P_m.[123] Array (a) performs far best because it enjoys resonance assistance (Scheme 13). The nonresonant groups in (b) and (c) are conformationally flexible allowing them to avoid formation of the hydrogen bond.

carboxylic acid groups in crystals donate a hydrogen bond to carboxylic acid acceptors (mostly forming the carboxylic acid dimer). The remaining 71% form hydrogen bonds with a great variety of other acceptors. A "relative success" of an alternative acceptor A competing with the carboxylic acid acceptor can be defined as $succ(A) = n(OH ··· A)/[n(OH ··· A) + n(OH ··· O_{carboxy})]$. This success rate was found to be over 90% for the strongest acceptors (COO⁻, P=O, N(Py), F⁻, Cl⁻). Water is also a very successful competitor ($succ(O_W) = 84\%$). The success rates are plotted against mean distances D for O acceptors in Figure 24. There is a clear correlation

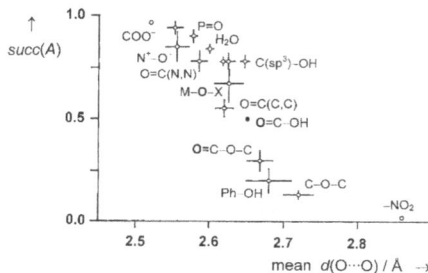

Figure 24. Correlation of the relative success of hydrogen bond acceptors competing for the carboxylic acid donor versus the mean hydrogen bond length.[124] Some important acceptors are identified as examples.

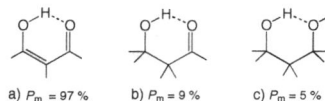

Scheme 17. Eight examples of intermolecular hydrogen bond motifs with their probability of formation (P_m) in crystals.[122] Notice that P_m of the carboxy–oxime heterodimer (b) is much higher than that of the carboxylic acid (e) and oxime homodimers (f).

motifs such as the carboxylic acid dimer can be explained by strong competition of alternative motifs. In a later study, the probability of formation was determined for 50 kinds of intramolecular hydrogen bonds.[123] The results for three topologically related rings with O–H ··· O hydrogen bonds are shown in Scheme 18. The probability of hydrogen bond formation for the β-diketone enolate (Scheme 18a) is close to 100% (because of RAHB), but it is very low for the nonresonant arrangements (Scheme 18b and c).

The formation of hydrogen bonds by the carboxylic acid donor group has been studied in detail.[124] Only 29% of all

between these quantities, and it should be noticed that even weak acceptors such as C–O–C (13%) and –NO₂ (2%) have a significant chance of attracting the strong carboxylic acid donor. Engineering the carboxylic acid dimer (Scheme 7) clearly requires the absence of successful competitors. If, for example, a pyridyl-N atom is present as a competitor, it is much more likely that a carboxylic acid–pyridine dimer is formed ($succ(N_{Py}) = 91\%$) than a carboxylic acid dimer (Scheme 19).

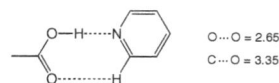

Scheme 19. Hydrogen-bonded dimer of a carboxylic acid and a pyridyl group; note the C–H ··· O interaction that is formed in addition to the (much stronger) O–H ··· N hydrogen bond. Mean geometries found in a CSD analysis ($n=32$) are for O–H ··· O: $d = 1.68(1)$, $D = 2.65(1)$ Å, $\theta = 171(1)°$; for C–H ··· O: $d = 2.53(3)$, $D = 3.35(2)$ Å, $\theta = 127(1)°$.[10c]

REVIEWS _____ T. Steiner

Generally, one makes the observation that the probability of formation increases with the number of hydrogen bonds constituting a motif. Whereas "two-point recognition" normally operates only moderately well, three-point recognition is clearly better, and four-point recognition is highly successful.[125]

7. Very Strong Hydrogen Bonds

Unlike moderate and weak hydrogen bonds, strong hydrogen bonds are quasi-covalent in nature,[28] and deserve special discussion. If the hydrogen bond is understood as an incipient proton-transfer reaction, a moderate hydrogen bond represents an early stage of such a reaction, while a strong one represents an advanced stage. It may be noted that from this viewpoint, hydrogen bonds with mainly ionic and little covalent nature are *not* classified as "strong" despite high dissociation energies ($-NH_3^+ \cdots Cl^-$ etc.). Instead, they might be termed ionic interactions with a moderate hydrogen bond formed on top.

Research on strong hydrogen bonds was pioneered by spectroscopists carrying out vibrational studies in solution,[44, 104, 126] and was only recently rediscovered by structural chemists. A key finding of spectroscopy is that very strong hydrogen bonds are formed only if the pK_a values of the partners are suitably matching. If the pK_a values are very different, either a moderate $X-H \cdots Y$ or an ionic $X^- \cdots H-Y^+$ hydrogen bond is formed, both of which are not very covalent. The quasi-covalent situation occurs in a certain "critical" range of ΔpK_a, the numerical characteristics of which depend on the particular system (for tabualtions, see ref. [126]). The pK_a value is a solution property that is not even defined in crystals, and these relationships (and related ones on matching of proton affinities) can not be transferred to the solid state in a general way. A polar environment favors polar forms of a hydrogen bond, and a hydrogen bond that is quasi-covalent in CCl_4 solution may become ionic in a polar crystal field. In an apolar crystal field, on the other hand, it may possibly remain in a state similar to apolar solution.

In this context, substances are interesting which have an apolar molecular periphery and a single hydrogen bond that is more or less buried inside. The crystal and solution properties of the hydrogen bond in these systems may be surprisingly similar. One example are adducts of pentachlorophenol with pyridines, for which the ΔpK_a value also defines the protonation state in crystals (Figure 25).[104] Strong hydrogen bonds can be grouped into several classes. The combination of acids with their conjugate base is well known (Scheme 20).[127] An exact matching of the ΔpK_a values is clear here, and proton transfer leads to a chemically equivalent situation ($X-H \cdots X^- \rightleftarrows X^- \cdots H-X$ or $X^+-H \cdots X \rightleftarrows X \cdots H-X^+$). Gilli and co-workers call these categories "negative" and "positive charge assisted hydrogen bonds", and compiled numerous examples with structural and, in part, also spectroscopic properties.[28, 46, 116] Strong hydrogen bonds of these types form easily, both inter- as well as intramolecularly. Examples of intramolecular hydrogen bonds of this

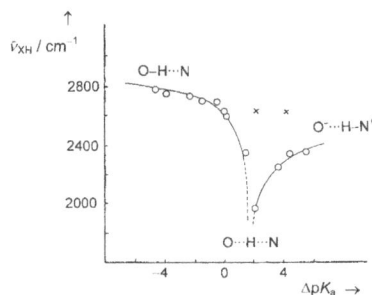

Figure 25. Correlation of ΔpK_a and the IR stretching frequency of crystalline pentachlorophenol–amine adducts. The symbols $\times$ indicate outliers, presumably caused by crystal-field effects. Very strong hydrogen bonds occur only in a certain "critical" region of ΔpK_a values.[104a]

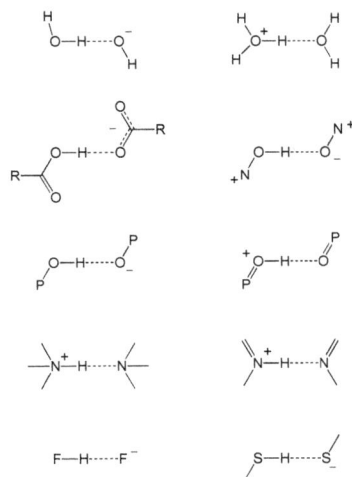

Scheme 20. Examples of strong hydrogen bonds between acids and complementary bases.

kind occur in hydrogenmaleate and hydrogenphthalate anions,[128] and in the cations of proton sponges.[129]

In systems with resonance-assisted hydrogen bonding (RAHB), such as in the β-diketone enols, proton transfer also leads to a symmetrical situation (Figure 22). RAHB thus provides a mechanism for the formation of strong hydrogen bonds in uncharged systems.[28] Substituents attached to the carbon skeleton may disturb the chemical symmetry and thereby weaken the hydrogen bond. For an accurate case study on a particular molecule, the reader is referred to work on the structure of nitromalonamide.[130]

Chemically unrelated partners can form strong hydrogen bonds only, if their pK_a values match at least roughly. Interesting examples are hydrogen bonds between carboxylic acids and N-oxides. These systems can be tuned by suitable chemical substitutions to form molecular adducts of the type $O=C-O-H \cdots O^- -N^+$, ionic adducts $-CO_2^- \cdots H-O-N^+$, or

Angew. Chem. Int. Ed. **2002**, *41*, 48–76

the intermediate quasi-covalent type with the H atom placed midway between two O atoms. The latter situation is found intra- and intermolecularly (Scheme 21), with $O \cdots O$ distances almost as short as the shortest possible value of 2.39 Å.[131, 132]

a)　　　　　　　　　　　　　　　b)

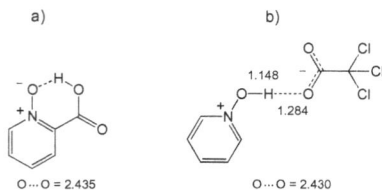

Scheme 21. Very short $O-H-O$ hydrogen bonds between carboxylic acid and $N-O$ groups. a) Intramolecular hydrogen bond in picolinic acid N-oxide.[131] b) Intermolecular hydrogen bond in the adduct formed between pyridine-N-oxide and trichloroacetic acid.[132] (Numerical values are obtained from a neutron diffraction crystal structure.[132b])

Not too many systems are known with strong and geometrically symmetric hydrogen bonds between different atom types. Even though the spectroscopic data predicts very short $N-H-O$ hydrogen bonds for a number of species,[126] the first example in a crystalline solid was characterized by neutron diffraction only very recently in the adduct of 4-methylpyridine and pentachlorophenol (Figure 26).[93] Even very small chemical changes in the system lead to loss of the symmetry, and to formation of either molecular $O-H \cdots N$ or ionic $O^- \cdots H-N^+$ hydrogen bonds.[133]

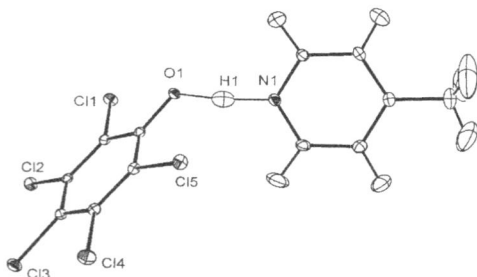

Figure 26. First example of an $O-H-N$ hydrogen bond with a centered position of the proton: pentachlorophenol/4-methylpyridine at 100 K characterized by neutron diffraction studies ($O \cdots N = 2.51$ Å).[93]

8. Weak and Less Common Hydrogen Bonds

In recent years, weak and less common types of hydrogen bonds have been a major topic in hydrogen bond research. Several reviews and a comprehensive book summarize this work,[8] and it is not necessary to review the field here once more in detail. The following remarks are intended to guide the interested reader through the large volume of modern literature on this topic.

Weak hydrogen bonds with $C-H$ groups as donors are studied best. Formerly considered "unusual" or "nonconventional", they are now discussed rather frequently in most fields of structural chemistry and biology. Dissociation energies are $0.4-4$ kcal mol^{-1}, with the majority < 2 kcal mol^{-1}. At the low energy end of the range, the $C-H \cdots O$ hydrogen bond gradually fades into a van der Waals interaction. The strong end of the interaction has not yet been well explored; $C-H \cdots A$ bonds stronger than 4 kcal mol^{-1} can readily be predicted to occur when very acidic $C-H$ or very basic acceptor groups are involved. Many properties of $C-H \cdots O/N/Hal$ interactions have been discussed in previous sections (Tables $1-3$, Figures 2, 8, 23; Schemes 10, 19). $C-H \cdots O$ hydrogen bonding in biological structures has also been investigated intensively.[134] Several recent reviews on the field are available.[81, 135–137]

A second type of weak hydrogen bond that has become well established in recent years are hydrogen bonds with π acceptors ($\pi=$Ph, C≡C, C=C, Py, Im, etc.; see Tables $1-3$, Figure 10). The bond energies for strong O/N−H donors are higher than for $C-H \cdots A$ bonds, about $2-4$ kcal mol^{-1} with neutral donors, and over 15 kcal mol^{-1} with charged donors. In the latter case, the $X-H \cdots \pi$ bond lies in the borderline region with the cation–π interaction (see Section 2.6). $X-H \cdots \pi$ hydrogen bonds occur in many fields of structural chemistry and biology, and there is a number of recent reviews that make a detailed discussion here unnecessary.[8, 27k, 138] A representative example is provided by an $N-H \cdots \pi$ bond stabilizing a 3_{10}-turn of a protein as shown in Figure 27.[80]

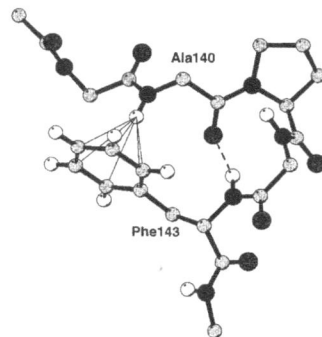

Figure 27. Example of an $N-H \cdots \pi$(Ph) hydrogen bond in a protein (stabilizing a 3_{10}-turn). Many related hydrogen bonds have been found that stabilize secondary structure elements.[80]

$X-H \cdots \pi$ interactions with weakly polar donors such as $C-H$ groups are interesting because $C-H$ groups span a wide range of polarities. With acidic donors such as C≡C−H, the $C-H \cdots \pi$ interaction manifests itself through its spectroscopic properties as a "true" hydrogen bond,[139] and calculated energies are in the $2-3$ kcal mol^{-1} range (Table 1). A particularly interesting group is C≡C−H, because it can act as a donor and an acceptor simultaneously. This situation may lead to the formation of chains and rings, C≡C−H $\cdots$ C≡C−H $\cdots$ C≡C−H or $O-H \cdots$ C≡C−H $\cdots O-H$, which are topologically equivalent to hydrogen bond chains involving only O–H groups.[8, 50, 139, 140]

The C–H $\cdots \pi$ interaction keeps structure-determining ability as the C–H polarity decreases, but it becomes a matter of debate whether the term "hydrogen bond" is still adequate.[8] According to the definition proposed in Section 2.1, it is justified as long as some proton donor properties of C–H can be detected. Even if this definition is interpreted liberally, (Ph)C–H $\cdots$ Ph and –CH$_3$ $\cdots$ Ph interactions belong to the borderline region of hydrogen bonding. The phenomenon of the "CH/π interaction" in organic and biological chemistry has been explored and reviewed by Nishio et al. in a very interesting way.[141, 142] S–H groups have also been shown to donate hydrogen bonds to π acceptors, but bond energies seem to be very low (<1 kcal mol^{-1}).[143]

Hydrogen bonds donated by metal hydrides, M–H $\cdots$ A, and accepted by electron-rich transition metal atoms, X–H $\cdots$ M, have been found in organometallic compounds.[8, 68, 144] The "dihydrogen bond" is formed between a protic X–H and a hydridic H–Y group, X$^{\delta-}$–H$^{\delta+}$ $\cdots$ H$^{\delta-}$–Y$^{\delta+}$, such as between N–H and H–B.[145] Many other special kinds of hydrogen bonds have been documented structurally,[8] and investigated theoretically.[146] The role of organic fluorine as a hydrogen bond acceptor has attracted particular interest.[27q, 67, 147]

Finally, X–H groups with reverse polarity, X$^{\delta+}$–H$^{\delta-}$, need to be mentioned. They can form directional interactions X$^{\delta+}$–H$^{\delta-}$ $\cdots$ A$^{\delta+}$ that are in many aspects analogous to a hydrogen bond. Since X–H does not act as proton donor, the term "hydrogen bond" as defined in Section 2.1 is not appropriate here. The alternative term "inverse hydrogen bond" suggested in a theoretical study reflects the chemical situation quite nicely.[148]

9. Bond Valence Concept

The idea that there is a more or less strict relationship between bond length and "bond order" or "valence" s dates back to Pauling.[3] Several expressions for $s = f(d)$ have been proposed,[149, 150] but still the most popular one is Pauling's exponential relationship [Eq. (1)], where d_0 is the length of a

$$s = \exp[(d_0 - d)/b] \qquad (1)$$

single bond with $s = 1$ and b is a constant (typically around 0.37 Å).[149] The rule of bond order conservation requires that the sum of bond orders (Σs) for each atom type is constant in all bonding situations (for example, $\Sigma s = 4$ for C, $\Sigma s = 2$ for O, etc.).

In a hydrogen bond X–H $\cdots$ A, there are two chemical bonds, X–H and H $\cdots$ A, and their bond orders must add up to 1, that is, $s_{XH} + s_{HA} = 1$. Together with Equation (1), a relationship between the X–H and H $\cdots$ A bond lengths follows. For the homonuclear case, X = A, it is expressed by Equation (2):

$$r_{XH} = r_0 - b \ln\{1 - \exp[(r_0 - r_{HX})/b]\} \qquad (2)$$

In multifurcated hydrogen bonds (Scheme 2) the bond orders of all H $\cdots$ A contacts must be included in the sum

$\Sigma s = 1$. The parameters r_0 and b can be obtained by fitting Equation (2) to experimental structural data,[31a] such as those shown in Figure 14. The most recent values of $r_0 = 0.928$ Å and $b = 0.393$,[36, 92] provide a good fit for O–H $\cdots$ O hydrogen bonds over the whole distance range (solid line in Figure 14a; there are slight but systematic deviations in the long distance range that could be remedied only by using a more sophisticated model function). Calculated bond orders s are given for O–H and H $\cdots$ O bonds over a wide distance range in Table 7. It may seem surprising that the valence of H $\cdots$ O in a hydrogen bond with $d = 1.8$ Å is as high as 0.11, whereas the one of O–H is reduced to as little as 0.89. This could be interpreted as a substantial covalent contribution even to moderate hydrogen bonds. The covalent contribution to the moderate hydrogen bonds in ice has recently been studied by Compton scattering methods.[151]

Table 7. Bond orders of O–H and H $\cdots$ O bonds as calculated from Equation (1) using the parameters $r_{0,OH} = 0.927$ Å, $b_{OH} = 0.395$ Å.[92]

O–H	s	H $\cdots$ O	s
0.97	0.90	1.20	0.50
0.98	0.87	1.30	0.39
0.99	0.85	1.40	0.30
1.0	0.83	1.50	0.23
1.02	0.79	1.60	0.18
1.04	0.75	1.70	0.14
1.06	0.71	1.80	0.11
1.08	0.69	1.90	0.08
1.10	0.65	2.00	0.07
1.15	0.57	2.20	0.04
1.20	0.50	2.40	0.02

Much less experimental material is available for N–H $\cdots$ N hydrogen bonds. In particular, the central region of the plot corresponding to Figure 14a is completely devoid of data points. The parameters $r_{0,NH} = 0.996$ Å and $b_{NH} = 0.381$ are much less reliable, and should not be expected to represent the region of very short hydrogen bonds very accurately.[94] The parameters for heteronuclear hydrogen bonds are also of limited accuracy because of small numbers of data.[36]

The bond valence concept can be used to rationalize effects of hydrogen bonding on non-hydrogen molecular frameworks, and reasonable numerical estimates can often be obtained. For example, the rule of bond order conservation requires that the O=C bond in C–O–H $\cdots$ O=C is weakened by the amount s_{HO}, and is lengthened correspondingly.[152]

For covalent bonds, it has been shown that s it is proportional to the electron density at the bond critical point, ρ_{BCP}.[39a] In O–H $\cdots$ O hydrogen bonds, both ρ_{BCP} and s depend in an exponential way on the H $\cdots$ O distance [Figure 3 and Equation (1), respectively], which implies the relationship $s \sim \rho_{BCP}$, at least approximately.

10. Summary and Outlook

The present article has given an overview of hydrogen bonding in the solid state, with a focus on the structural properties. It is the mere volume of the published literature

       REVIEWS

that requires this article to concentrate on the fundamentals of the structural aspects. It was shown that the hydrogen bond phenomenon is a very broad one: there are dozens of different X–H $\cdots$ A interactions that occur commonly in the condensed phases, and in addition there are innumerable less common ones. Dissociation energies span more than two orders of magnitude, about $0.2-40$ kcal mol^{-1}. Within this range, the nature of the interaction is not constant, but includes electrostatic, covalent, and dispersion contributions in varying weights. To further increase the complexity, the hydrogen bond has broad transition regions—"grey areas"—with the covalent bond, the van der Waals interaction, the ionic interaction, and also the cation–π interaction.

Looking into the future of a complex subject is always difficult, but a little history sometimes helps. The field of hydrogen bonds did not evolve smoothly, but saw periods of rapid development and periods of stagnation. In particular in the 1980s, many researchers felt that the field was more or less closed. Around 1990, the field openened again, and it is still moving rapidly. The present boom is related to the broad spectrum of fields that are involved: material science, inorganic and organic chemistry, biology, and pharmacy. In none of these areas has the role and function of hydrogen bonds been explored satisfactorily, nor has hydrogen bond research reached a level that allows the interaction to be *controlled* (with the exception of some aspects of "crystal engineering"). With this background, it is not too daring to predict that gaining control of hydrogen bonds, and developing corresponding tools to be utilized in the fields mentioned above will be the next goal.

I would like to thank the many co-authors of my original publications who have helped me gain insight into the fascinating topic of hydrogen bonding. There are too many names to be mentioned, over 50 in total, so I may only name a few individually: G. R. Desiraju, G. Koellner, J. Kroon, W. Saenger, and J. L. Sussman. Over the years, various sources have provided financial support, and I would like to thank in particular the Deutsche Forschungsgemeinschaft and the Minerva Foundation. A referee of this review has provided an extraordinary number of helpful suggestions for improvement. I would like to thank this anonymous colleague for the very careful reading and the large amount of time spent.

Received: March 6, 2001 [A456]

[1] The discovery of the hydrogen bond cannot be attributed to a single author, and no genuine "first paper" can be quoted. Specialized articles developing relevant ideas began to appear at the beginning of the 20th century, mainly in the German and English literature, but the far-reaching relevance of the hydrogen bond was not yet recognized. More elaborate studies and clear general concepts were published from the 1920s on, with pioneering roles usually attributed to Latimer and Rodebush, Huggins, and Pauling. By the end of the 1930s, a "classical" view of the hydrogen bond was established that dominated the field for half a century. Research into hydrogen bonds experienced a peak in the 1950s and 1960s, followed by relative stagnation from the mid-1970s to the late 1980s. An intense revival occurred from about 1990 on. Theoretical concepts were long dominated by Coulsons partitioning into resonance forms, which were later replaced by modern quantum chemical models. Historical surveys can be found in the books cited below.[2–9]

[2] G. C. Pimentel, A. L. McClellan, *The Hydrogen Bond*, Freeman, San Francisco, **1960**.

[3] L. Pauling, *The Nature of the Chemical Bond*, 3rd ed., Cornell University Press, Ithaca, **1963**.

[4] W. C. Hamilton, J. A. Ibers, *Hydrogen Bonding in Solids*, Benjamin, New York, **1968**.

[5] G. A. Jeffrey, W. Saenger, *Hydrogen Bonding in Biological Structures*, Springer, Berlin, **1991**.

[6] G. A. Jeffrey, *An Introduction to Hydrogen Bonding*, Oxford University Press, Oxford, **1997**.

[7] S. Scheiner, *Hydrogen Bonding. A Theoretical Perspective*, Oxford University Press, Oxford, **1997**.

[8] G. R. Desiraju, T. Steiner, *The Weak Hydrogen Bond in Structural Chemistry and Biology*, Oxford University Press, Oxford, **1999**.

[9] Monographs: a) *The Hydrogen Bond* (Ed.: D. Hadzi), Pergamon, New York, **1957**; b) *The Hydrogen Bond. Recent Developments in Theory and Experiment*, Vols. *1 – 3* (Eds.: P. Schuster, G. Zundel, C. Sandorfy), North-Holland, Amsterdam, **1976**; c) *Theoretical Treatments of Hydrogen Bonding* (Ed.: D. Hadzi), Wiley, Chichester, **1997**.

[10] Database information retrieved for this article was obtained from the Cambridge Structural Database (CSD: F. H. Allen, O. Kennard, *Chem. Des. Autom. News.* **1993**, *8*, 1), update 5.20 (Oct. 2000) with 224 400 entries, ordered and error-free organic and organometallic crystal structures with *R* values < 0.06, X–H bond lengths normalized. Standard uncertainties of mean values were calculated only if $n > 5$. Geometric cutoff criteria were used as specified: a) Tables 3, 5, 6: for X = O, N: $d < 2.4$ Å; for X = S, C: $d < 2.8$ Å, $\theta > 135°$. b) For Table 4: $d(O/N/F^-/F–X^-/F–TM) < 2.4$ Å, $d(S/Se/Cl^-/Cl–X^-/Cl–TM) < 2.8$ Å, $d(Br^-/Br–TM) < 3.0$ Å, $d(I^-/I–TM/Cl–C) < 3.2$ Å, $d(\pi) < 3.0$ Å where π is the π-bond centroid (TM = transition metal); angle θ in all cases > 135°. c) Schemes 4 and 19: $d < 3.0$ Å, $\theta > 90°$. d) Scheme 6: $d < 2.4$ Å, $\theta > 90°$, $n = 26$. e) Scheme 7: neutron diffraction data, $R < 0.08$, $d < 2.4$ Å, $n = 7$. Standard uncertainties are around 0.003 Å and 0.4° for values not involving the H atom. f) Figure 20: only 4-Me-Py acceptors.

[11] Still the most popular source of van der Waals radii is an article by A. Bondi, *J. Phys. Chem.* **1964**, *68*, 441 – 451, who gives values of 1.2 Å for H, 1.52 Å for O, 1.55 Å for N, 1.70 Å for C, 1.75 Å for Cl, and 1.80 Å for S. The sum of van der Waals radii is then, for example, 2.72 Å for H $\cdots$ O. In a recent re-investigation based on crystal structure data, very similar values were obtained with the exception of the H atom which is found to be slightly smaller with a radius of 1.1 Å (R. S. Rowland, T. Taylor, *J. Phys. Chem.* **1996**, *100*, 7384 – 7391).

[12] G. R. Desiraju, *Angew. Chem.* **1995**, *107*, 2541 – 2558; *Angew. Chem. Int. Ed. Engl.* **1995**, *34*, 2311 – 2327.

[13] a) G. R. Desiraju, *Crystal Engineering. The Design of Organic Solids*, Elsevier, Amsterdam, **1989**; b) G. R. Desiraju, *Chem. Commun.* **1997**, 1475 – 1482.

[14] The term "crystal engineering" was coined by Gerhard Schmidt at the Weizmann Institute of Science. He worked on the construction of crystal packing schemes with the aim of aligning reactive groups in crystals in such a way that they are ready for topochemical reactions. For a review, see G. M. J. Schmidt, *Pure Appl. Chem.* **1971**, *27*, 647 – 678.

[15] a) L. Leiserowitz, *Acta Crystallogr. Sect. B* **1976**, *32*, 775 – 802; b) Z. Berkovitch-Yellin, L. Leiserowitz, *J. Am. Chem. Soc.* **1982**, *104*, 4052 – 4064; c) J. Bernstein, M. C. Etter, L. Leiserowitz in *Structure Correlation*, Vol. 2 (Eds.: H.-B. Bürgi, J. D. Dunitz), VCH, Weinheim, **1994**, pp. 431 – 507; d) I. Weissbuch, R. Popovitz-Biro, M. Lahav, L. Leiserowitz, *Acta Crystallogr. Sect. B* **1995**, *51*, 115 – 148.

[16] a) J. C. McDonald, G. M. Whitesides, *Chem. Rev.* **1994**, *94*, 2383 – 2420; b) C. B. Aakeröy, *Acta Crystallogr. Sect. B* **1997**, *53*, 569 – 583; c) L. R. McGillivray, J. L. Atwood, *Angew. Chem.* **1998**, *110*, 1029 – 1031; *Angew. Chem. Int. Ed.* **1998**, *38*, 1019 – 1034; d) M. J. Zaworotko, *Chem. Commun.* **2001**, 1 – 9.

[17] L. N. Kuleshova, P. M. Zorkii, *Acta Crystallogr. Sect. B* **1980**, *36*, 2113 – 2115.

[18] a) M. C. Etter, *Acc. Chem. Res.* **1990**, *23*, 120 – 126; b) M. C. Etter, J. C. MacDonald, J. Bernstein, *Acta Crystallogr. Sect. B* **1990**, 256 – 262; c) W. D. S. Motherwell, G. P. Shields, F. H. Allen, *Acta Crystallogr. Sect. B* **1999**, *55*, 1044 – 1056.

[19] J. Bernstein, R. E. Davis, L. Shimoni, N.-L. Chang, *Angew. Chem.* **1995**, *107*, 1689 – 1708; *Angew. Chem. Int. Ed. Engl.* **1995**, *34*, 1555 – 1573.

[20] a) A. K. Rappé, E. R. Bernstein, *J. Phys. Chem. A* **2000**, *104*, 6117 – 6128; b) K. Müller-Dethlefs, P. Hobza, *Chem. Rev.* **2000**, *100*, 143 – 167; c) M. J. Calhorda, *Chem. Commun.* **2000**, 801 – 809.

[21] The so-called van der Waals cutoff-definition of the hydrogen bond (van der Waals is not guilty of it) requires that the H ··· A (or the X ··· A) separation is shorter than the sum of the van der Waals radii. For X–H ··· O contacts, this is about H ··· O < 2.6 – 2.7 Å.[11] At this distance, the interaction is thought to be switched from "hydrogen bond" to "van der Waals" type. This definition is faulty and severely misleading:[5, 6] the electrostatic field of dipoles (or full charges, if ions are involved) does not terminate at any cutoff distance, and the hydrogen bond is experimentally detectable at distances far beyond the van der Waals contact distance.[8] The understanding of the van der Waals cutoff criterion in the context of hydrogen bonding is of historical relevance for science in general. It is one of the examples where a concept, without experimental support, can persist for decades in the literature and to some degree rise to the level of a dogma. When reading the larger part of the literature on hydrogen bonds written in the period of about 1970 to 1990, the concept is uncritically adhered to as the "natural" hydrogen bond definition, and disregards the fact that no experimental proof or theoretical evidence of it had been available. For a few years now, the definition has been gradually disappearing from the literature.

[22] T. Steiner, G. R. Desiraju, *Chem. Commun.* **1998**, 891 – 892.

[23] J. D. Dunitz, A. Gavezzotti, *Acc. Chem. Res.* **1999**, *32*, 677 – 684.

[24] K. Morokuma, *Acc. Chem. Res.* **1977**, *10*, 294 – 300.

[25] A. E. Reed, L. A. Curtiss, F. Weinhold, *Chem. Rev.* **1988**, *88*, 899 – 926.

[26] D. S. Coombes, S. L. Price, D. J. Willock, M. Leslie, *J. Phys. Chem.* **1996**, *100*, 7352 – 7360.

[27] a) S. Gronert, *J. Am. Chem. Soc.* **1993**, *115*, 10258 – 10266; b) J. E. Del Bene, M. J. Frisch, J. A. Pople, *J. Phys. Chem.* **1988**, *89*, 3669 – 3674; c) J. E. Del Bene, *J. Phys. Chem.* **1988**, *92*, 2874 – 2880; d) W.-L. Zhu, X.-J. Tan, C. M. Puah, J.-D. Gu, H.-L. Jiang, K.-X. Chen, C. E. Felder, I. Silman, J. L. Sussman, *J. Phys. Chem. A* **2000**, *104*, 9573 – 9580; e) T. Neuheuser, B. A. Hess, C. Reutel, E. Weber, *J. Phys. Chem.* **1994**, *98*, 6459 – 6467; f) Y. Gu, T. Kar, S. Scheiner, *J. Am. Chem. Soc.* **1999**, *121*, 9411 – 9422; g) M. W. Feyereisen, D. Feller, D. A. Dixon, *J. Phys. Chem.* **1996**, *100*, 2993 – 2997; h) L. Turi, J. J. Dannenberg, *J. Phys. Chem.* **1993**, *97*, 7899 – 7909; i) S. Tsuzuki, K. Honda, T. Uchimaru, M. Mikami, K. Tanabe, *J. Am. Chem. Soc.* **2000**, *122*, 11450 – 11458; j) I. Alkorta, S. Maluendes, *J. Phys. Chem.* **1995**, *99*, 6457 – 6460; k) J. F. Malone, C. M. Murray, M. H. Charlton, R. Docherty, A. J. Lavery, *J. Chem. Soc. Faraday Trans.* **1997**, *93*, 3429 – 3436; l) J. J. Novoa, F. Mota, *Chem. Phys. Lett.* **2000**, *318*, 345 – 354; m) E. L. Woodbridge, T.-L. Tso, M. P. McGrath, W. J. Hehre, E. K. C. Lee, *J. Chem. Phys.* **1986**, *85*, 6991 – 6994; n) T. van Mourik, F. B. van Duijneveldt, *J. Mol. Struct.* **1995**, *341*, 63 – 73; o) J. J. Novoa, B. Tarron, M. H. Whangbo, J. M. Williams, *J. Chem. Phys.* **1991**, *95*, 5179 – 5186; p) M. M. Szczesniak, G. Chalasinski, S. M. Cybulski, P. Cieplak, *J. Chem. Phys.* **1993**, *98*, 3078 – 3089; q) J. A. K. Howard, V. J. Hoy, D. O'Hagan, G. T. Smith, *Tetrahedron* **1996**, *52*, 12613 – 12622.

[28] P. Gilli, V. Ferretti, V. Bertolasi, G. Gilli, *J. Am. Chem. Soc.* **1994**, *116*, 909 – 915.

[29] J. C. Ma, D. A. Dougherty, *Chem. Rev.* **1997**, *97*, 1303 – 132.

[30] T. Steiner, S. A. Mason, *Acta Crystallogr. Sect. B* **2000**, *56*, 254 – 260.

[31] a) H.-B. Bürgi, *Angew. Chem.* **1975**, *87*, 461 – 475; *Angew. Chem. Int. Ed. Engl.* **1975**, *14*, 460 – 474; b) H.-B. Bürgi, J. D. Dunitz, *Acc. Chem. Res.* **1983**, *16*, 153 – 161.

[32] F. A. Cotton, R. L. Luck, *Inorg. Chem.* **1989**, *28*, 3210 – 3213.

[33] C. B. Aakeroy, K. R. Seddon, *Chem. Soc. Rev.* **1993**, *2*, 397 – 407.

[34] F. H. Allen, *Acta Crystallogr. Sect. B* **1986**, *42*, 515 – 522.

[35] G. A. Jeffrey, L. Lewis, *Carbohydr. Res.* **1978**, *60*, 179 – 182.

[36] T. Steiner, *J. Phys. Chem. A* **1998**, *102*, 7041 – 7052.

[37] P. Coppens, *X-Ray Charge Densities and Chemical Bonding*, Oxford University Press, Oxford, **1997**.

[38] Nothing like a complete literature survey of charge density studies in the field of hydrogen bonds can be given here. For a small selection of

recent examples, see a) P. R. Mallison, K. Wozniak, C. C. Wilson, K. L. McCormack, D. S. Yufit, *J. Am. Chem. Soc.* **1999**, *121*, 4640 – 4646; b) W. D. Arnold, L. K. Sanders, M. T. McMahon, A. V. Volkov, G. Wu, P. Coppens, S. R. Wilson, N. Godbout, E. Oldfield, *J. Am. Chem. Soc.* **2000**, *122*, 4708 – 4717; c) P. Macchi, B. Iversen, A. Sironi, B. C. Chakoumakos, F. K. Larsen, *Angew. Chem.* **2000**, *112*, 2831 – 2834; *Angew. Chem. Int. Ed.* **2000**, *39*, 2719 – 2722; d) R. S. Gopalan, P. Kumaradhas, G. U. Kulkarni, C. R. N. Rao, *J. Mol. Struct.* **2000**, *521*, 97 – 106.

[39] a) R. F. W. Bader, *Atoms in Molecules. A Quantum Theory*, Oxford University Press, Oxford, **1990**; b) R. F. W. Bader, P. L. A. Popelier, T. A. Keith, *Angew. Chem.* **1994**, *106*, 647 – 659; *Angew. Chem. Int. Ed. Engl.* **1994**, *33*, 620 – 631; c) R. G. A. Bone, R. F. W. Bader, *J. Phys. Chem.* **1996**, *100*, 10892 – 10911; d) R. F. W. Bader, *J. Phys. Chem. A* **1998**, *102*, 7314 – 7323; e) J. Hernandez-Trujillo, R. F. W. Bader, *J. Phys. Chem. A* **2000**, *104*, 1779 – 1794.

[40] U. Koch, P. L. A. Popelier, *J. Phys. Chem.* **1995**, *99*, 9747 – 9754.

[41] I. Alkorta, J. Elguero, *J. Phys. Chem. A* **1999**, *103*, 272 – 279.

[42] P. L. A. Popelier, *J. Phys. Chem. A* **1998**, *102*, 1873 – 1878.

[43] P. L. A. Popelier, G. Logothetis, *J. Organomet. Chem.* **1998**, *555*, 101 – 111.

[44] D. Hadzi, S. Bratos in *The Hydrogen Bond. Recent Developments in Theory and Experiment, Vol. 2* (Eds.: P. Schuster, G. Zundel, C. Sandorfy), North Holland, Amsterdam, **1976**, pp. 565 – 612.

[45] a) A. Novak, *Struct. Bonding* **1974**, *18*, 177 – 216; b) E. Libowitzky, *Monatsh. Chem.* **1999**, *130*, 1047 – 1059.

[46] G. Gilli, P. Gilli, *J. Mol. Struct.* **2000**, *552*, 1 – 15.

[47] a) G. R. Desiraju, B. N. Murty, *Chem. Phys. Lett.* **1987**, *139*, 360 – 361; b) B. T. G. Lutz, J. H. van der Maas, J. A. Kanters, *J. Mol. Struct.* **1994**, *325*, 203 – 214; c) see ref. [8], p. 42, for a more recent diagram.

[48] S. Borycka, M. Rozenberg, A. M. M. Schreurs, J. Kroon, E. B. Starikov, T. Steiner, *New J. Chem.* **2001**, *25*, 1111 – 1113.

[49] B. Lutz, J. A. Kanters, J. van der Maas, J. Kroon, T. Steiner, *J. Mol. Struct.* **1998**, *440*, 81 – 87.

[50] An interestig model case is the crystalline ethynyl steroid mestranol, which contains two molecules in the asymmetric unit. Despite very similar C≡C–H ··· O hydrogen bond lengths (H ··· O = 2.44 and 2.48 Å), the $\Delta \bar{\nu}_{CH}$ values of the two ethynyl groups differ by a factor of almost three (21.3 and 59.8 cm^{-1}). This difference is explained by different roles of the ethynyl group in the heterogeneous hydrogen bond network: one is part of a coorperative array, and the other is part of an anticooperative one (T. Steiner, B. Lutz, J. van der Maas, N. Veldman, A. M. M. Schreurs, J. Kroon, J. A. Kanters, *Chem. Commun.* **1997**, 191 – 192).

[51] M. Rozenberg, A. Loewenschuss, Y. Marcus, *Phys. Chem. Chem. Phys.* **2000**, 2699 – 2673.

[52] B. T. G. Lutz, J. Jacob, J. H. van der Maas, *Vib. Spectrosc.* **1996**, *12*, 197 – 206.

[53] A. V. Iogansen, *Spectrochim. Acta A* **1999**, *55*, 1585 – 1612.

[54] A. Allerhand, P. von R. Schleyer, *J. Am. Chem. Soc.* **1963**, *85*, 1715 – 1723.

[55] a) P. Hobza, Z. Havlas, *Chem. Rev.* **2000**, *100*, 4253 – 4264; b) P. Hobza, *Phys. Chem. Chem. Phys.* **2001**, *3*, 2555 – 2556.

[56] Correlations of ^{1}H chemical shift with O ··· O bond length have been set up by different authors for chemically different systems. The results are essentially always the same, and show universal validity of the correlation. Examples can be found, for example, in: a) for hydrated silica glasses: H. Eckert, J. P. Yesinowski, L. A. Silver, E. M. Stolper, *J. Phys. Chem.* **1988**, *92*, 2055 – 2064; b) for a set of organic molecules with intramolecular resonance assisted hydrogen bonds: V. Bertolasi, P. Gilli, V. Ferretti, G. Gilli, *J. Chem. Soc. Perkin Trans. 2* **1997**, 945 – 952; c) for a set of biological solids: T. K. Harris, Q. Zhao, A. S. Mildvan, *J. Mol. Struct.* **2000**, *552*, 97 – 109.

[57] S. N. Smirnov, N. S. Golubev, G. S. Denisov, H. Benedict, P. Schah-Mohammedi, H.-H. Limbach, *J. Am. Chem. Soc.* **1996**, *118*, 4094 – 4101.

[58] Hydrogen bonds can be classified into three strength categories in different ways, that is, with demarcations between the categories placed differently, and different names can be attached to the categories. In the literature, one can find sets of names such as "very strong, strong, weak", "strong, moderate, weak", and "strong, weak, very weak". Clearly, hydrogen bonds between, for example, water

Angew. Chem. Int. Ed. **2002**, *41*, 48 – 76

molecules, are quite "strong" for one researcher and fairly "weak" for the other one, depending on the personal focus of interest. In a general view on hydrogen bonds, and essentially following the categorization of Jeffrey,[6] it seems appropriate to attach the names "strong" and "weak" to the extremes of the scale, and use a term such as "moderate" for the intermediate range. One might note that chemically, the difference between "strong" (quasi-covalent nature) and "moderate" (mainly electrostatic) is larger than between "moderate" (electrostatic) and "weak" (electrostatic/dispersion).

[59] J. Kroon, J. A. Kanters, *Nature* **1974**, *248*, 667 – 669.
[60] T. Steiner, *Acta Crystallogr. Sect. B* **1998**, *54*, 456 – 463.
[61] T. Steiner, W. Saenger, *Acta Crystallogr. Sect. B* **1992**, *48*, 819 – 827.
[62] T. Steiner, W. Saenger, *J. Am. Chem. Soc.* **1992**, *114*, 10146 – 10154.
[63] I. Olovsson, P.-G. Jönsson in *The Hydrogen Bond. Recent Developments in Theory and Experiment, Vol. 2* (Eds.: P. Schuster, G. Zundel, C. Sandorfy), North Holland, Amsterdam, **1976**, pp. 393 – 456.
[64] H. F. J. Savage, J. Finney, *Nature* **1986**, *322*, 717 – 720.
[65] D. Braga, F. Grepioni, E. Tedesco, *Organometallics* **1998**, *17*, 2669 – 2672.
[66] The interpretation of *d* – θ scatter plots of C–H ··· Cl–C interactions is somewhat controversial: a) C. B. Aakeröy, T. A. Evans, K. R. Seddon, I. Pálinko, *New J. Chem.* **1999**, *23*, 145 – 152; b) P. K. Thallapally, A. Nangia, *CrystEngComm* **2001**, 27.
[67] V. R. Thalladi, H.-C. Weiss, D. Bläser, R. Boese, A. Nangia, G. R. Desiraju, *J. Am. Chem. Soc.* **1998**, *120*, 8702 – 8710.
[68] L. Brammer, D. Zhao, F. T. Lapido, J. Braddock-Wilking, *Acta Crystallogr. Sect. B* **1995**, *51*, 632 – 640.
[69] F. H. Allen, C. M. Bird, R. S. Rowland, P. R. Raithby, *Acta Crystallogr. Sect. B* **1997**, *53*, 680 – 695.
[70] T. Steiner, J. A. Kanters, J. Kroon, *Chem. Commun.* **1996**, 1277 – 1278.
[71] J. Kroon, J. A. Kanters, J. C. G. M. van Duijneveldt-van de Rijdt, F. B. van Duijneveldt, J. A. Vliegenthart, *J. Mol. Struct.* **1975**, *24*, 109 – 129.
[72] A. L. Llamas-Saiz, C. Foces-Foces, O. Mo, M. Yanez, J. Elguero, *Acta Crystallogr. Sect. B* **1992**, *48*, 700 – 713.
[73] G. Yap, A. L. Rheingold, P. Das, R. H. Crabtree, *Inorg. Chem.* **1995**, *34*, 3474 – 3476.
[74] G. Aullón, D. Bellamy, L. Brammer, E. Bruton, A. G. Orpen, *Chem. Commun.* **1998**, 653 – 654.
[75] L. Brammer, E. A. Bruton, P. Sherwood, *New J. Chem.* **1999**, *23*, 965 – 968.
[76] G. A. Worth, R. C. Wade, *J. Phys. Chem.* **1995**, *99*, 17473 – 17482.
[77] T. Steiner, A. M. M. Schreurs, M. Lutz, J. Kroon, *New J. Chem.* **2001**, *25*, 174 – 178.
[78] T. Steiner, S. A. Mason, M. Tamm, *Acta Crystallogr. Sect. B* **1997**, *53*, 843 – 848.
[79] S. Suzuki, P. G. Green, R. E. Bumgarner, S. Dasgupta, W. A. Goddard III, G. A. Blake, *Science* **1992**, *257*, 942 – 945.
[80] T. Steiner, G. Koellner, *J. Mol. Biol.* **2001**, *305*, 535 – 557.
[81] T. Steiner, *Chem. Commun.* **1997**, 727 – 734.
[82] Of the distributions shown in Figure 11, only Figure 11a can be characterized properly by statistical descriptors such as mean value, standard deviation, and 90 %-range. The distribution shown in Figure 11b has a defined peak, but if the fraction of multifurcated bonds is large, most other descriptors are more or less useless. The numerical mean value, for example, will strongly depend on the distance cutoff chosen, and will furthermore be sample-dependent. To obtain numerical data of general relevance, it is therefore justified to concentrate on fairly linear hydrogen bonds as in Figure 11a, and keep in mind that minor components of bifurcated bonds may be formed in addition. This situation corresponds to selecting the densely populated clusters of Figure 7, and only analyze these.
[83] T. Steiner, *New J. Chem.* **1998**, *22*, 1099 – 1103.
[84] Y. Umezawa, S. Tsuboyama, K. Honda, J. Uzawa, M. Nishio, *Bull. Chem. Soc. Jpn.* **1998**, *71*, 1207 – 1213.
[85] R. Preißner, U. Egner, W. Saenger, *FEBS Lett.* **1991**, *288*, 192 – 196.
[86] I. Rozas, I. Alkorta, J. Elguero, *J. Phys. Chem. A* **1998**, *102*, 9925 – 9932.
[87] T. Steiner, S. A. Mason, *Z. Kristallogr. Suppl.* **2001**, *18*, 93.
[88] a) H. S. Rzepa, M. H. Smith, M. L. Webb, *J. Chem. Soc. Perkin Trans. 2* **1994**, 703 – 707; b) M. Pilkington, J. D. Wallis, S. Larsen, *J. Chem.*

Soc. Chem. Commun. **1995**, 1499 – 1500; c) P. K. Bakshi, A. Linden, B. R. Vincent, S. P. Roe, D. Adhikesavalu, T. S. Cameron, O. Knop, *Can. J. Chem.* **1994**, *72*, 1273 – 1293; d) K.-T. Wei, D. L. Ward, *Acta Crystallogr. Sect. B* **1976**, *32*, 2768 – 2773.
[89] W. H. Baur, *Acta Crystallogr. Sect. B* **1972**, *28*, 1456 – 1465.
[90] T. Steiner, W. Saenger, *Acta Crystallogr. Sect. B* **1991**, *47*, 1022 – 1023.
[91] K. Nakamoto, M. Margoshes, R. E. Rundle, *J. Am. Chem. Soc.* **1955**, *77*, 6480 – 6486.
[92] T. Steiner, W. Saenger, *Acta Crystallogr. Sect. B* **1994**, *50*, 348 – 357.
[93] T. Steiner, I. Majerz, C. C. Wilson, *Angew. Chem.* **2001**, *113*, 2728 – 2731; *Angew. Chem. Int. Ed.* **2001**, *40*, 2651 – 2654.
[94] T. Steiner, *J. Chem. Soc. Chem. Commun.* **1995**, 1331 – 1332.
[95] T. Steiner, *J. Chem. Soc. Perkin Trans. 2* **1995**, 1315 – 1319.
[96] P. L. A. Popelier, R. F. W. Bader, *Chem. Phys. Lett.* **1992**, *189*, 542 – 548.
[97] T. Steiner, *Acta Crystallogr. Sect. B* **1998**, *54*, 464 – 470.
[98] T. Steiner, *J. Phys. Chem. A* **2000**, *104*, 433 – 435.
[99] P. Seiler, R. Weisman, E. D. Glendening, F. Weinhold, V. B. Johnson, J. D. Dunitz, *Angew. Chem.* **1987**, *99*, 1216 – 1218; *Angew. Chem. Int. Ed. Engl.* **1987**, *26*, 1175 – 1177.
[100] I. Chao, J.-C. Chen, *Angew. Chem.* **1996**, *108*, 200 – 202; *Angew. Chem. Int. Ed. Engl.* **1996**, *35*, 195 – 197.
[101] J. A. Kanters, J. Kroon, R. Hooft, A. Schouten, J. A. M. van Schijndel, J. Brandsen, *Croat. Chim. Acta* **1991**, *64*, 353 – 370.
[102] T. Steiner, *Chem. Commun.* **1999**, 2299 – 2300.
[103] I. Majerz, Z. Malarski, L. Sobczyk, *Chem. Phys. Lett.* **1997**, *274*, 361 – 364.
[104] a) Z. Malarski, M. Rospenk, L. Sobczyk, E. Grech, *J. Phys. Chem.* **1982**, *86*, 401 – 406; b) L. Sobczyk, *Ber. Bunsenges. Phys. Chem.* **1998**, *102*, 377 – 383.
[105] D. Mootz, H.-G. Wussow, *J. Chem. Phys.* **1981**, *75*, 1517 – 1522.
[106] D. Mootz, J. Hocken, *Z. Naturforsch. B* **1989**, *44*, 1239 – 1246.
[107] A. R. Ubbelohde, K. J. Gallagher, *Acta Crystallogr.* **1955**, *8*, 71 – 83.
[108] M. Ichikawa, *J. Mol. Struct.* **2000**, *552*, 63 – 70.
[109] D. Mootz, M. Schilling, *J. Am. Chem. Soc.* **1992**, *114*, 7435 – 7439.
[110] G. A. Jeffrey, *Crystallogr. Rev.* **1995**, *3*, 213 – 260.
[111] G. A. Jeffrey, J. Mitra, *Acta Crystallogr. Sect. B* **1983**, *39*, 469 – 480.
[112] T. Steiner, S. A. Mason, W. Saenger, *J. Am. Chem. Soc.* **1991**, *113*, 5676 – 5687.
[113] B. M. Kariuki, K. D. M. Harris, D. Philp, J. M. A. Robinson, *J. Am. Chem. Soc.* **1997**, *119*, 12679 – 12670.
[114] G. Gilli, F. Bellucci, V. Ferretti, V. Bertolasi, *J. Am. Chem. Soc.* **1989**, *111*, 1023 – 1028.
[115] V. Bertolasi, P. Gilli, V. Verretti, G. Gilli, *Chem. Eur. J.* **1996**, *2*, 925 – 934.
[116] a) V. Bertolasi, P. Gilli, V. Ferretti, G. Gilli, *Acta Crystallogr. Sect. B* **1995**, *51*, 1004 – 1015; b) P. Gilli, V. Bertolasi, V. Ferretti, G. Gilli, *J. Am. Chem. Soc.* **2000**, *122*, 10205 – 10217.
[117] T. Steiner, *Chem. Commun.* **1998**, 411 – 412.
[118] B. Krebs, *Angew. Chem.* **1983**, *95*, 113 – 134; *Angew. Chem. Int. Ed. Engl.* **1983**, *22*, 113 – 134.
[119] W. Saenger, *Nature* **1979**, *279*, 343 – 344.
[120] a) I. D. Brown, *Acta Crystallogr. Sect. A* **1976**, *32*, 24 – 31; b) P. Murray-Rust, J. P. Glusker, *J. Am. Chem. Soc.* **1984**, *106*, 1018 – 1025; c) R. O. Gould, A. M. Gray, P. Taylor, M. D. Walkinshaw, *J. Am. Chem. Soc.* **1985**, *107*, 5921 – 5927; d) V. R. Bartenev, N. G. Kameneva, A. A. Lipanov, *Acta Crystallogr. Sect. B* **1987**, *43*, 275 – 280; e) B. Lesyng, G. A. Jeffrey, H. Maluszynska, *Acta Crystallogr. Sect. B* **1988**, *44*, 193 – 198.
[121] H.-B. Bürgi, J. D. Dunitz, *Acta Crystallogr. Sect. B* **1988**, *44*, 445 – 448.
[122] F. H. Allen, W. D. S. Motherwell, P. R. Raithby, G. P. Shields, R. Taylor, *New J. Chem.* **1999**, *23*, 25 – 34.
[123] C. Bilton, F. H. Allen, G. P. Shields, J. A. K. Howard, *Acta Crystallogr. Sect. B* **2000**, *56*, 849 – 856.
[124] T. Steiner, *Acta Crystallogr. Sect. B* **2001**, *57*, 103 – 106.
[125] F. H. Beijer, H. Kooijman, A. L. Spek, R. P. Sijbesma, E. W. Meijer, *Angew. Chem.* **1998**, *110*, 79 – 82; *Angew. Chem. Int. Ed.* **1998**, *37*, 75 – 78; minireview of four hydrogen bonding motif: C. Schmuck, W. Wienand, *Angew. Chem.* **2001**, *113*, 4493 – 4499; *Angew. Chem. Int. Ed.* **2001**, *40*, 4363 – 4369.
[126] G. Zundel, *Adv. Chem. Phys.* **2000**, *111*, 1 – 217.
[127] F. Hibbert, J. Emsley, *Adv. Phys. Org. Chem.* **1990**, *26*, 255 – 379.

REVIEWS T. Steiner

[128] F. Fillaux, N. Leygue, J. Tomkinson, A. Cousson, W. Paulus, *Chem. Phys.* **1999**, *244*, 387–403; H. Küppers, F. Takusagawa, T. F. Koetzle, *J. Chem. Phys.* **1985**, *82*, 5636–5647.

[129] a) A. L. Llamas-Saiz, C. Foces-Foces, J. Elguero, *J. Mol. Struct.* **1994**, *328*, 297–323; b) H. A. Staab, C. Krieger, G. Hieber, K. Oberdorf, *Angew. Chem.* **1997**, *109*, 1946–1949; *Angew. Chem. Int. Ed. Engl.* **1997**, *36*, 1884–1886.

[130] G. K. H. Madsen, C. Wilson, T. M. Nymand, G. J. McIntyre, F. K. Larsen, *J. Phys. Chem. A* **1999**, *103*, 8684–8690.

[131] a) T. Steiner, A. M. M. Schreurs, M. Lutz, J. Kroon, *Acta Crystallogr. Sect. C* **2000**, *56*, 577–579; b) J. Stare, J. Mavri, G. Ambrozic, D. Hadzi, *THEOCHEM* **2000**, *500*, 429–440.

[132] a) L. Golic, D. Hadzi, F. Lazarini, *J. Chem. Soc. Chem. Commun.* **1971**, 860–861; b) K. D. Eichhorn, *Z. Kristallogr.* **1991**, *195*, 205–220; c) T. Steiner, A. M. M. Schreurs, J. Kroon, D. Hadzi, unpublished results.

[133] a) T. Steiner, C. C. Wilson, I. Majerz, *Chem. Commun.* **2000**, 1231–1232; b) T. Steiner, S. A. Mason, C. C. Wilson, I. Majerz, *Z. Kristallogr. Suppl.* **2000**, 68.

[134] a) Z. S. Derewenda, L. Lee, U. Derewenda, *J. Mol. Biol.* **1995**, *252*, 248–262; b) J. Bella, H. Berman, *J. Mol. Biol.* **1996**, *264*, 734–742; c) G. F. Fabiola, S. Krishnaswamy, V. Nagarajan, V. Patthabi, *Acta Crystallogr. Sect. D* **1997**, *53*, 316–312; d) P. Chakrabarti, S. Chakrabarti, *J. Mol. Biol.* **1998**, *284*, 867–873; e) C. Bon, M. S. Lehmann, C. Wilkinson, *Acta Crystallogr. Sect. D* **1999**, *55*, 978–987; f) R. Vargas, J. Garza, D. A. Dixon, B. P. Hay, *J. Am. Chem. Soc.* **2000**, *122*, 4750–4755.

[135] a) G. R. Desiraju, *Acc. Chem. Res.* **1996**, *29*, 441–449; b) G. R. Desiraju, *Acc. Chem. Res.* **1991**, *24*, 270–276.

[136] T. Steiner, *Crystallogr. Rev.* **1996**, *6*, 1–57.

[137] M. Wahl, M. Sundaralingam, *Trends Biochem. Sci.* **1997**, *22*, 97–102.

[138] M. F. Perutz, *Philos. Trans. R. Soc. London Ser. A* **1993**, *345*, 105–112.

[139] T. Steiner, E. B. Starikov, A. M. Amado, J. J. C. Teixeira-Dias, *J. Chem. Soc. Perkin Trans. 2* **1995**, 1321–1326.

[140] a) T. Steiner, *Adv. Mol. Struct. Res.* **1998**, *4*, 43–77; b) T. Steiner, M. Tamm, B. Lutz, J. van der Maas, *Chem. Commun.* **1996**, 1127–1128; c) K. Subramanian, S. Lakshmi, K. Rajagopalan, G. Koellner, T. Steiner, *J. Mol. Struct.* **1996**, *384*, 121–126; d) M. A. Viswamitra, R. Radhakrishnan, J. Bandekar, G. R. Desiraju, *J. Am. Chem. Soc.* **1993**, *115*, 4868–4869.

[141] a) M. Nishio, M. Hirota, *Tetrahedron* **1989**, *45*, 7201–7245; b) M. Nishio, Y. Umezawa, M. Hirota, Y. Takeuchi, *Tetrahedron* **1995**, *51*, 8665–8701.

[142] M. Nishio, M. Hirota, Y. Umezawa, *The CH/π Interaction. Evidence, Nature and Consequences*, Wiley, New York, **1998**.

[143] M. S. Rozenberg, T. Nishio, T. Steiner, *New J. Chem.* **1999**, *23*, 585–586.

[144] D. Braga, F. Grepioni, G. R. Desiraju, *Chem. Rev.* **1998**, *98*, 1375–1405.

[145] R. H. Crabtree, P. E. M. Siegbahn, O. Eisenstein, A. L. Rheingold, T. F. Koetzle, *Acc. Chem. Res.* **1996**, *29*, 348–354.

[146] I. Alkorta, I. Rozas, J. Elguero, *Chem. Soc. Rev.* **1998**, *27*, 163–179.

[147] J. D. Dunitz, R. Taylor, *Chem. Eur. J.* **1997**, *3*, 89–98.

[148] I. Rozas, I. Alkorta, J. Elguero, *J. Phys. Chem. A* **1997**, *101*, 4236–4244.

[149] I. D. Brown, *Acta Crystallogr. Sect. B* **1992**, *48*, 553–572.

[150] F. Mohri, *Acta Crystallogr. Sect. B* **2000**, *56*, 626–638.

[151] E. D. Isaacs, A. Shulka, P. M. Platzman, D. R. Hamann, B. Barbellini, C. A. Rulk, *Phys. Rev. Lett.* **1999**, *82*, 600–603. However, see also the criticism by A. H. Romero, P. L. Silvestrelli, M. Parrinello, *Phys. Status Solidi B* **2000**, *220*, 703–708, and by T. K. Ghanty, V. N. Staroverov, P. R. Koren, E. R. Davidson, *J. Am. Chem. Soc.* **2000**, *122*, 1210–1214, and references therein.

[152] S. J. Grabowski, *Tetrahedron* **1998**, *54*, 10153–10160.

Name Index

Subject Index

www.ingramcontent.com/pod-product-compliance
Lightning Source LLC
Chambersburg PA
CBHW081401270326
41930CB00015B/3381